T0205492

Financial Mathematics and Fintech

Series Editors

Zhiyong Zheng, Renmin University of China, Beijing, Beijing, China

Alan Peng, University of Toronto, Toronto, ON, Canada

This series addresses the emerging advances in mathematical theory related to finance and application research from all the fintech perspectives. It is a series of monographs and contributed volumes focusing on the in-depth exploration of financial mathematics such as applied mathematics, statistics, optimization, and scientific computation, and fintech applications such as artificial intelligence, block chain, cloud computing, and big data. This series is featured by the comprehensive understanding and practical application of financial mathematics and fintech. This book series involves cutting-edge applications of financial mathematics and fintech in practical programs and companies.

The Financial Mathematics and Fintech book series promotes the exchange of emerging theory and technology of financial mathematics and fintech between academia and financial practitioner. It aims to provide a timely reflection of the state of art in mathematics and computer science facing to the application of finance. As a collection, this book series provides valuable resources to a wide audience in academia, the finance community, government employees related to finance and anyone else looking to expand their knowledge in financial mathematics and fintech.

The key words in this series include but are not limited to:

a) Financial mathematics
b) Fintech
c) Computer science
d) Artificial intelligence
e) Big data

Thomas Barrau • Raphael Douady

Artificial Intelligence for Financial Markets

The Polymodel Approach

 Springer

Thomas Barrau
AXA Investment Managers Chorus Ltd
Hong Kong, Hong Kong S.A.R.

Raphael Douady
Economic Center
University Paris 1 Sorbonne
Paris, France

ISSN 2662-7167   ISSN 2662-7175   (electronic)
Financial Mathematics and Fintech
ISBN 978-3-030-97321-6   ISBN 978-3-030-97319-3   (eBook)
https://doi.org/10.1007/978-3-030-97319-3

JEL Classification: C49; C58; G01; G11; G17
Mathematics Subject Classification: 62P20; 62P25; 68T01; 91B05; 91G10; 91G15; 91G70

This Springer imprint is published by the registered company Springer Nature Switzerland AG
The registered company address is: Gewerbestrasse 11, 6330 Cham, Switzerland

Foreword

This book exposes a manageable approach to exploit nonlinearity for the purpose of financial investment. It summarizes the research Raphael Douady and co-authors have conducted for nearly two decades now. This research is expanded on multiple aspects by a joint work conducted by Thomas Barrau.

In general, relationships between variables are nonlinear and noisy. When attempting to relate a target variable to several explanatory variables, modeling finely multi-dimensional linearity suffers from the curse of dimensionality without bringing blatant benefits. Polymodels simplify this approach by retaining unidimensional nonlinearities and proposing various aggregation methods according to the task at hand. For instance, in the case of prediction, each one-dimensional model produces a prediction. The most relevant ones are then selected, then possibly averaged in case of ambiguity, to yield the final prediction. In the case of defining a risk indicator, as downturns are usually associated with an increase of correlations, the distribution of goodness-of-fit metrics is analyzed. It can be seen as a form of ensemble learning.

The first task is to perform the one-dimensional nonlinear regressions. Several methods are presented, including non-parametric kernel regressions but the method of choice is linear regression on nonlinear functions (Hermite polynomials for their orthogonality properties) of the explanatory variable, further regularized by shrinkage to the linear component to mitigate overfitting.

Nonlinear transforms of initial factors can be used as new variables or features to feed a machine learning algorithm. Another use of extracted nonlinearities is to transform their convexity into an antifragility score (in the sense of N.N. Taleb). These scores can then be used as a factor to build long-short portfolios. The Polymodel approach can be implemented at the level of the market, industries, or individual stocks. With the help of Bayesian conditioning it can combine the information of hundreds of factors.

In summary, polymodels provide a very workable methodology to exploit nonlinearity for the purpose of investing, forecasting and risk management. Barrau and Douady describe the method and its application in this book filled with examples and backtests which will certainly inspire numerous finance practitioners.

Quantitative Research, Bloomberg LP   Bruno Dupire
New York, NY, USA

University of New York
New York, NY, USA

Preface

Like many innovative ideas, *polymodels* emerged from a very stressful situation. Soon after we started the Riskdata venture, a new risk software aimed at the buy-side, the Twin Towers in New York were hit. This event would change the face of the world, and seriously put our nascent project at stake. There was no chance to sign a single contract for at least a year. While we were about to abandon the project and find a payroll job, a fund of hedge fund manager proposed the following challenge to us: *given* only *audited monthly returns of the funds, what can we reliably say about their strategy, their risks, and how they fit in a portfolio, made of both liquid and alternative assets?*

This challenge was way beyond the mere assessment of classical performance metrics, such as the Sharpe ratio, downside volatility, beta, Fama–French factor analysis, etc. These measures, widely used by the industry, are known to provide relevant information from the past, but nothing truly reliable for the future behavior. We figured out that, if a software, however sophisticated, ignores the key points addressed in a conversation between an investor and a manager, it would miss the elephant in the room. So, we asked this manager to let us silently attend some of his meetings with the managers he was either invested in or planning to. To our surprise, the investor would spend only a few minutes talking about correlations and volatility, then the rest of the hour on specific moments that were relevant to the particular strategy: how did the manager weather such or such an event?; when and how did he flip his positions?; and other such questions. Our challenge was to put these questions into a software that was systematic enough to automatically address them. The idea was *not* to strictly reproduce the questions, as the answers would not explicitly be in the fund returns, but to create a statistical framework that would, mechanically, provide the best answers, given the returns information.

The solution had to incorporate the financial environment in which the fund evolves. A return, in itself, is not highly informative. It only becomes so when related to the rest of the market. The first question to address is: which market factors are statistically related to the fund, and how? Note that here, the range of possible factors can be extremely broad if we don't want to miss major risks. A multilinear regression on just a few factors would definitely not address the issue. The second observation is that unusually large events bear most of the information. Extrapolating from the correlation of small events is limited and prone to surprises from threshold effects.

Then another aspect of the challenge popped up: how can we make such a specific analysis comparable across the various funds, in order to get a full vision of the portfolio risks? The portfolio may contain hundreds of investment lines. Its risk can be due to the blow up of one particular investment with too large a weight, or, more severely, to the unexpected joint loss of a large portion of the portfolio, which we could call *vanishing diversification* upon some particular event.

The solution came in the form of a thorough factor analysis of all the funds in the portfolio, with two important features:

1. Each risk factor should be tested against *all* the funds in the portfolio.
2. Large events should be emphasized to anticipate predictable correlation breaks.

The *polymodel* approach naturally emerged as the only possible solution to this problem. Traditional multifactor analysis first seeks a set of relevant factors for each fund, then evaluates the importance of each factor and its "coefficient of impact" (the *factor loadings*). Correlation breaks and change of regimes are separately addressed, often in a very rudimentary manner, and rarely in their systematicity. The *polymodel* approach, on the contrary, first looks at the impact of each single factor on all the components of the portfolio, taking into account systematic correlation breaks through a non-linear model, then, and only then, aggregates the single factor information to answer such questions as the global portfolio risks or the contribution of a particular line in the portfolio.

One can view this analysis as a gigantic automatic generator of stress tests. Instead of characterizing each fund and the entire portfolio as a proxy made of a combination of risk factors, it represents each asset, each fund, and the portfolio itself, by its list of responses to all these stress tests, each being identified by a factor shift. Effects of vanishing diversification, for instance, are particularly well anticipated, since we know that, for instance, under a sudden surge in interest rates, all asset classes will be impacted: equities, credit spreads, etc.

We are aware that no statistical method escapes the fact that it is based on the past and that the future has surprises in store for us. But the room left for surprises by *polymodels* is definitely much narrower than that left by traditional linear multifactor analysis.

After successfully applying the *polymodel* approach to portfolios of hedge funds, major crises, such as the 2007 Subprime Crisis and the 2008 Global Financial Crisis, made it obvious that this non-linear approach was relevant beyond the hedge funds universe, to the entire investment space: liquid investments (equities, fixed-income, derivatives. . .), private equity, real estate, etc. We also figured out that, beyond a simple risk/return analysis, the richness of *polymodels* allowed us to answer a much broader range of questions, such as the anticipated behavior of a manager in a liquidity crisis, strategy and style drift, the impact of macroeconomic conditions on financial markets, crisis prediction, etc. This book contains methods to address several of these questions.

Changes of regime and correlation breaks systematically occur upon large events. In our view, a sound mathematical model must include any observed systematic phenomenon. In short, the mathematics should fit reality, and not the opposite. In this regard, like any other artificial intelligence method, *polymodels* first seek to bring effective solutions to empirical problems.

The organization of the present book naturally results from this aim, as it is structured around the idea of solving an empirical problem. As the most generic problem of finance is the prediction of the financial markets, this is the issue that we chose to approach.

Therefore, we propose a simple model of the returns of a portfolio invested in the stock market, decomposed into market, industry and specific returns. We make predictions of these three different components using various techniques, all based on non-linear modeling done with polymodels. Empirical applications are complemented by a theoretical introduction to Polymodel Theory, and we propose pragmatic solutions to the various challenges it raises. We finally aggregate the predictions through a genetic algorithm, and the predictive power of the combined forecast is assessed by the simulation of a trading strategy. The portfolio obtained reaches more than twice the Sharpe ratio of the market, after transaction costs.

Although the primary aim of the present book was to support a doctoral thesis, it became clear during its redaction that its interest was far beyond this purpose. Being oriented to deliver applications while being simultaneously solidly grounded by a self-contained theoretical presentation of *polymodels*, the book is a turnkey manual to approach financial markets with artificial intelligence. It is a suitable handbook for the practitioner, as it proposes novel and concrete solutions to solve real-world problems. Thus, the book will be of interest to machine learners searching for new prediction algorithms, quantitative traders looking for original strategies, risk managers trying to avoid drawdowns, and portfolio managers dealing with the question of the combination of several trading strategies.

We would like to acknowledge several people for their contributions to the book.

Charles-Albert Lehalle and Yuri Kabanov provided an excellent review of the entire book, allowing improvements on various statistical and machine learning aspects.
Augustin Landier provided smart perspectives on questions related to factor investing, and more generally about financial economics.
Jean-Philippe Bouchaud, Jean-Paul Laurent and Alexander Lipton proposed some clever improvements to the manuscript during its first public presentation.
Nassim Taleb, the Quantitative Finance team of the Stony Brook University of New York (especially Robert Frey and Andrew Mullhaupt), the Riskdata research team, Igmar Adlerberg, Bertrand Cabrit, Iliya Zovko, Cyril Coste, the fund of hedge fund managers who trusted us, Bernad Lozé, Jeffrey Tarrant, as well as Adil Abdulali, Bruno Dupire, Marco Avellaneda, Emmanuel Derman, Dominique Guégan, Shivaji Rao, all the members of the Real-World Risk Institute, and Hector Chan, Gavin Chan, Olivier Parent-Colombel, Pierre-Emmanuel Juillard from AXA IM Chorus shared fruitful discussions and insights.

Aurelien Vallée and Omar Masmoudi greatly helped in the development of the code
     required for the various applications of the book, and in dealing with large
     clusters of machines to perform computations.
Finally, we acknowledge the Sorbonne Center of Economics (CES), for its material
     support and for the discussions we had with its members.

Hong Kong, Hong Kong S.A.R.   Thomas Barrau
Paris, France   Raphael Douady

Contents

Chapter 1
Introduction

Abstract As polymodels are introduced as an artificial intelligence technique for the prediction of financial markets, we begin the book by presenting a simple framework to produce these predictions. We start with a concise literature review on the prediction of stock returns. A simple model is then presented, and we explain how contributing to the model actually coincides with contributing to the literature. We finally develop the plan of the book, which answers each of the different points evocated about the literature of financial market prediction.

Keywords Polymodel theory · Prediction framework · Modeling · Stock market · Financial markets

1.1 Financial Market Predictions: A Concise Literature Review

The possibility of making predictions of financial markets was first suggested by Bachelier (1900), who used a random walk to model the price changes of commodities and stocks. The modeling of financial markets following a random walk implies the validity of the efficient market hypothesis (see Samuelson, 1965), which states that asset prices reflect all available information, leading to the impossibility of predicting price changes (Fama, 1965). These seminal works, undisputed at the time, are now partly invalidated by a large literature coming from different fields such as economics, applied mathematics, and the econophysics of psychology, which tackle the question of the prediction of financial markets from various different angles.

Although there may be several approaches to pursuing the goal of "beating the market" consistently, perhaps the most obvious one is to simply predict the returns of a given stock index.

Economists proposed to use various macroeconomic variables as predictors of the market returns. Although their studies may suffer from some form of overfitting, Rapach et al. (2010) showed that combinations of the forecasts made by these variables make it possible to consistently predict the market. Furthermore, their results are based on out-of-sample data, thus pointing to the structural stability of the macroeconomic roots of the equity premium. In a following study (Neely et al., 2014), this macroeconomic approach is completed by the use of technical indicators based on trend, momentum and volume.

Others focused on the analysis of extreme market events: the prediction of crashes. Following a physics-inspired approach,[1] Sornette developed a model based on the identification of a power law trend in the prices, inevitably leading to an unsustainable bubble, and thus, a crash (Johansen et al., 2000; Sornette, 2009). Market crashes have also been predicted using a variety of correlation-based methods (see for example Patro et al., 2013; Zheng et al., 2012; Douady & Kornprobst, 2018; Harmon et al., 2011).[2]

All these prediction methods, whether they are concentrated on predicting common or extreme events, even though they use very different approaches, share a common aim: providing reliable predictions about the future market returns.

However, even in-sample, the returns of the market provide an insufficient description of the variance of each stock return. It has been shown that the industry component also matters (King, 1966), and the specific component of the stock returns, i.e. the part that does not depend on the market or the industry, is even more important (Roll, 1988).

Just like the market returns, the industry returns have also been shown to be predictable, for example by exploiting the momentum anomaly, which states that past high (low) industry returns are followed by future high (low) industry returns (O'Neal, 2000). Since there are several industries, it is also possible to predict industry returns in a cross-sectional fashion, using industry characteristics (Asness et al., 2000). In this paper, Asness also demonstrates that it is possible to perform cross-sectional predictions of specific returns, however, most of the papers dedicated to the cross-section of stock returns do not control for industry exposures.

The most important milestone in this literature may be the seminal paper of Fama and French (1993), which introduces a three-factor linear model to explain the cross-section of stock returns. This model was completed in 2015 by the same two authors (Fama & French, 2015), leading to a five-factor model that takes into account the beta of the firm (its sensitivity to the market returns), its size (the market capitalization), a value factor (the book to market ratio), its (operating) profitability and its investment pattern (aggressive versus conservative investment behavior of the firm). Although it is not included in the model, the momentum anomaly is also an important explanatory factor of the cross-section of stock returns (see Carhart, 1997; Jegadeesh & Titman, 1993).

This approach of describing the cross-section of stock returns by a linear multi-factor model has been extended so much that there are now dozens to hundreds of factors recognized by the literature (see, for example, Green et al. (2014) and Hou et al. (2017) for an assessment of the reliability of the different factors). It also

---

[1] The application of physics in the field of economics is known as 'Econophysics'.

[2] Chapter 4 provides a more detailed description of these prediction methods.

shaped the asset management industry, in which the so-called "risk premia"[3] funds tend to emerge as a new asset class, investing in a passive manner into these factors (Bender et al., 2014). What used to be called "alpha", in the sense of Jensen (1968), is now simply considered as risk factors, making these strategies comparable to "beta" investing, i.e. passive investing in the market. Following the approach of predicting ("timing") the market, a literature finally emerged on the question of timing the returns of the risk factors (see Bender et al., 2018; Asness et al., 2017).

1.2   A Simple Model of Portfolio Returns

Ultimately, although they rely on very different approaches, the works mentioned above all aim to produce expectations about the future stock returns. However, to the best of our knowledge at the time of writing, a single framework that combines all these kinds of predictions does not exist in the literature. We propose such a simple, descriptive model below. Note that that the goal of the model is efficiency. It does not aim to propose an economic explanation of the stock returns.

We start by returning to the distinction, proposed by Roll (1988), among the three main components of the stock return: a market component, an industry component and a firm-specific component.

This distinction can be formalized by the following formula:

$$r_a = \omega_a(r_M, r_I, r_S). \tag{1.1}$$

Here, r_a is the stock return of a given stock "a", while r_M, r_I, and r_S are respectively the market return, the industry return and the specific return for this stock. Note that the definition of the function $\omega_a()$ has been avoided at this stage for the sake of generality.

It is clear from the literature that the industry and the specific returns can be explained (respectively) by the cross-sections of industry and firm characteristics (size, value, quality, etc.). Following Fama and French (1993, 2015), for each characteristic we can form a portfolio, so that the stock returns are expressed as a function of the factors' portfolio returns.

For example, we can create a portfolio that invests in the momentum anomaly. This portfolio, if constructed long/short, would be long the past winners and short the past losers, and to some extent may neutralize the exposure to the market returns. Hereafter we refer to the returns of this momentum factor portfolio as "r_{mom}". Note that the portfolio can be constructed with the same stocks in order to invest on the cross-section of industry or specific returns only. Indeed, we can neutralize the industry exposure of a portfolio, or its specific exposure by an appropriate long/

---

[3] Among the economical justifications of a factor is the idea that investors are rewarded for a risk associated with investment in that factor.

short construction (see Asness et al., 2014) for an example of such a portfolio construction). Hence, we adopt the name "$r_{mom,I}$" for the returns of the portfolio which invests in the momentum anomaly through the cross-section of industries while being neutralized at the specific level. Similarly, "$r_{mom,S}$" designates the returns of the portfolio which invests in the momentum anomaly through the cross-section of specific returns while being neutralized at the industry level.

However, there are numerous anomalies identified in the literature, both at the industry and specific level. We define the industry and specific sets of *all* the relevant factors (known and unknown) as follows:

$$\mathcal{F}_I = \{r_{mom,I}, r_{size,I}, r_{value,I}, r_{quality,I}, \ldots\},$$
$$\mathcal{F}_S = \{r_{mom,S}, r_{size,S}, r_{value,S}, r_{quality,S}, \ldots\}. \tag{1.2}$$

Assuming that these factors are able to completely describe the industry and firm characteristics, we can put

$$r_I = \eta_I(\mathcal{F}_I) \tag{1.3}$$

and

$$r_S = \eta_S(\mathcal{F}_S), \tag{1.4}$$

in which $\mathcal{F}_I$ and $\mathcal{F}_S$ are combined using their associated functions η_I and η_S to form the industry and specific returns. Substituting these into Eq. (1.1) leads to

$$r_a = \omega_a(r_M, \eta_I(\mathcal{F}_I), \eta_S(\mathcal{F}_S)). \tag{1.5}$$

Note that if we represent each stock return as a combination of the returns of several factor portfolios, we can also represent a portfolio "*P*", composed of a total number of *b* stocks, as a combination of the factor portfolios. Hence, the returns of such a portfolio would be:

$$r_P = \sum_{a=1}^{b} \omega_a(r_M, \eta_I(\mathcal{F}_I), \eta_S(\mathcal{F}_S)) \times w_a, \tag{1.6}$$

in which w_a is the weight of the stock "*a*" in the portfolio. The weighted sum of individual combination functions ω_a being just a veil, masking that we ultimately invest in factor portfolios, we may reformulate the portfolio's returns as follows:

$$r_P = \Omega_P(r_M, \mathcal{F}_I, \mathcal{F}_S), \tag{1.7}$$

where Ω_P is the combination function of sub-portfolios used to construct the portfolio *P*. An appropriate definition of Ω_P would reflect to some extent the weighting (potentially non-linear) among factors that are captured by the combination functions ω_a, η_I and η_S.

Nevertheless, respecting the average stock decomposition would lead to a portfolio statically exposed to the returns of the different factors. Maximizing r_p would require a dynamic adjustment of the exposures according to some criterion in order to benefit from timing the factor performances. For example, we should go short the market portfolio if we have information from a market-timing strategy that the market is about to enter into a bearish period. The combination function, properly defined, would thus be able to identify the proper proportions of the factor portfolios as well as to time their performances.

We propose below a basic example of what the return of a portfolio invested in well-known factors would be, following a simple, linear specification of omega:

$$r_{simple\_portfolio} = \beta_M r_M + \beta_{mom} r_{mom} + \beta_{size} r_{size} + \beta_{value} r_{value}, \qquad (1.8)$$

in which the betas are some weighting coefficients, which can be obtained using a linear model from historical data, for example. Such a naïve portfolio construction is easily achievable nowadays by investing in "smart beta" products (Bender et al., 2014).

Following the framework we proposed, superior portfolio performances, which add value on top of the simple portfolio above, may come from three sources:

- Identifying unknown factors, known as "alphas", which explain industry or specific returns.
- Timing the performance of the sub-portfolios, through factor or market timing.
- Properly assembling the different sub-portfolios, i.e. properly defining the weighting in omega.

The present book adds to the literature of financial market predictions by proposing contributions to all three of these sources using polymodels.

1.3   Plan of the Book

We propose novel methods to predict the financial markets. As stated, these methods are principally based on Polymodel Theory, a non-linear framework of analysis classified as an artificial intelligence method. The book is composed of eight chapters: an introduction; two chapters devoted to the presentation of techniques; four chapters which use these techniques to produce applications; and a conclusion.

Since they have only recently appeared in the literature, polymodels are described in the second chapter of the book. While staying quite general, we explain the overall interest of the method, which allows one to reduce overfitting and simultaneously increase precision, compared to standard multivariate alternatives.

The third chapter is dedicated to the estimation method we use for the polymodels, called "Linear Non-Linear Mixed" (LNLM), which uses cross-validation to regularize the non-linear part of the model. We consider in this chapter the theoretical performance of the LNLM model through a large panel of simulations.

The fourth chapter describes an application of Polymodel Theory to market timing. We see in this chapter that a polymodel allows one to capture the correlation structure the market maintains with its economic environment in a non-trivial manner, leading to a predictive signal of the direction of market returns.

The fifth chapter considers the prediction of the industry returns. Although we do not directly use a polymodel, we still rely on non-linear modeling by LNLM to reveal that non-linearity is priced in the cross-section of industries, in the form of an antifragility factor. The factor is shown to be an investable alpha.

The sixth chapter is about the prediction of the specific returns. We propose direct predictions of the specific returns using polymodels and offer techniques to take into account the correlations of the predictors while aggregating the multiple predictions of the polymodel. The predictions we obtain are confirmed to be a significant signal.

In the seventh chapter we combine the signals which predict the market, industry and specific returns into a single portfolio. The combination is performed using another artificial intelligence method, namely a genetic algorithm, allowing us to maximize the returns of the final portfolio, net of transaction costs.

The concluding chapter summarizes the results.

References

Asness, C., Chandra, S., Ilmanen, A., & Israel, R. (2017). Contrarian factor timing is deceptively difficult. *The Journal of Portfolio Management, 43*(5), 72–87.

Asness, C. S., Frazzini, A., & Pedersen, L. H. (2014). Low-risk investing without industry bets. *Financial Analysts Journal, 70*(4), 24–41.

Asness, C. S., Porter, R. B., & Stevens, R. L. (2000). Predicting stock returns using industry-relative firm characteristics. Available at SSRN 213872.

Bachelier, L. (1900). Théorie de la spéculation. *Annales scientifiques de l'École normale supérieure, 17*, 21–86.

Bender, J., Hammond, P. B., & Mok, W. (2014). Can alpha be captured by risk premia? *The Journal of Portfolio Management, 40*(2), 18–29.

Bender, J., Sun, X., Thomas, R., & Zdorovtsov, V. (2018). The promises and pitfalls of factor timing. *The Journal of Portfolio Management, 44*(4), 79–92.

Carhart, M. M. (1997). On persistence in mutual fund performance. *The Journal of Finance, 52*(1), 57–82.

Douady, R., & Kornprobst, A. (2018). An empirical approach to financial crisis indicators based on random matrices. *International Journal of Theoretical and Applied Finance, 21*(03), 1850022.

Fama, E. F. (1965). The behavior of stock-market prices. *The Journal of Business, 38*(1), 34–105.

Fama, E. F., & French, K. R. (1993). Common risk factors in the returns on stocks and bonds. *Journal of Financial Economics, 33*(1), 3–56.

Fama, E. F., & French, K. R. (2015). A five-factor asset pricing model. *Journal of Financial Economics, 116*(1), 1–22.

Green, J., Hand, J., & Zhang, F. (2014). The remarkable multidimensionality in the cross-section of expected US stock returns. Available at SSRN, 2262374.

Harmon, D., de Aguiar, M. A., Chinellato, D. D., Braha, D., Epstein, I., & Bar-Yam, Y. (2011). Predicting economic market crises using measures of collective panic. Available at SSRN 1829224.

Hou, K., Xue, C., & Zhang, L. (2017). *Replicating anomalies* (No. w23394). National Bureau of Economic Research.

Jegadeesh, N., & Titman, S. (1993). Returns to buying winners and selling losers: Implications for stock market efficiency. *The Journal of Finance, 48*(1), 65–91.

Jensen, M. C. (1968). The performance of mutual funds in the period 1945–1964. *The Journal of Finance, 23*(2), 389–416.

Johansen, A., Ledoit, O., & Sornette, D. (2000). Crashes as critical points. *International Journal of Theoretical and Applied Finance, 3*, 219–255.

King, B. F. (1966). Market and industry factors in stock price behavior. *The Journal of Business, 39*(1), 139–190.

Neely, C. J., Rapach, D. E., Tu, J., & Zhou, G. (2014). Forecasting the equity risk premium: the role of technical indicators. *Management Science, 60*(7), 1772–1791.

O'Neal, E. S. (2000). Industry momentum and sector mutual funds. *Financial Analysts Journal, 56*(4), 37–49.

Patro, D. K., Qi, M., & Sun, X. (2013). A simple indicator of systemic risk. *Journal of Financial Stability, 9*(1), 105–116.

Rapach, D. E., Strauss, J. K., & Zhou, G. (2010). Out-of-sample equity premium prediction: Combination forecasts and links to the real economy. *The Review of Financial Studies, 23*(2), 821–862.

Roll, R. (1988). R2 [J]. *Journal of Finance, 43*(3), 541–566.

Samuelson, P. A. (1965). Proof that properly anticipated prices fluctuate randomly. *Management Review, 6*(2), 41–49.

Sornette, D. (2009). Dragon Kings, Black Swans and the Prediction of crisis. *International Journal of Terraspace Science and Engineering, 2*(1), 1–18.

Zheng, Z., Podobnik, B., Feng, L., & Li, B. (2012). Changes in cross-correlations as an indicator for systemic risk. *Scientific Reports, 2*, 888.

Chapter 2
Polymodel Theory: An Overview

Abstract We present Polymodel Theory, defining a polymodel as a *collection of non-linear univariate models*. A mathematical formulation as well as an epistemological foundation is presented. We explain how polymodels are, in several respects, a superior alternative to classical multivariate regressions estimated with OLS, Ridge and Stepwise techniques; we also present the limits of the method. Although it is a regression technique, we clarify how the polymodels framework is closer to artificial intelligence than traditional statistics.

Keywords Polymodel theory · Artificial intelligence · Machine learning · Univariate regression · Multivariate regression · Non-linear modeling · High dimension modeling · Overfitting

2.1 Introduction

Polymodels, understood as *a collection of non-linear univariate models*, were introduced in finance by Coste et al. (2010). In their paper, polymodels are used as a part of an overall procedure to predict hedge fund performance. The concept, albeit fairly general, is thus presented concisely, since the paper aims to focus on the results of its applications. The purpose of the current chapter is therefore to provide a more in-depth discussion on the theory of polymodels, in order to understand the pros and cons of this technique, along with the possibilities of the framework it offers.

The use of a collection of univariate models must be understood as an alternative to the use of a multivariate regression model. The interest of polymodels is thus explained extensively, from this perspective, in the current chapter. However, the way we approach modeling through polymodels somehow differs from the standard perspective of statistics, as it is closer to artificial intelligence. For the reader to follow the standpoint we propose, we need to go back to the question of the purpose of modeling.

For Aris (1994), "a system of equations, Σ, is said to be a model of the prototypical system, S, if it is formulated to express the laws of S and its solution is intended to represent some aspect of the behavior of S". This quite static definition is complemented by Davis et al. (2011), who propose a list of the purposes for which

models are constructed, including among others "to influence further experimenta-
tion or observation". This last goal is the key to understanding our position. As it has
been designed to tackle problems encountered in finance, Polymodel Theory belongs
to the field of applied mathematics, since our focus is on "mathematics that finds
applications outside of its own interest" (see Davis et al. (2011) again for this
definition). Polymodels are a tool developed not only to observe and represent
(some sub-parts of) the financial system, but also with the goal of acting as
practitioners, traders and risk managers that are a part of it, thus modifying the
system. Fundamentally, our approach to modeling is thus oriented by the pursuit of
effective results while acting in the real world.

This objective being stated, we suggest below a simplified, caricature version of
the modeling process. This representation is not intended to describe the process that
each researcher follows, nor to outline a methodological standard; it simply offers
some support to the presentation of our approach to modeling. We start with the
problem of having one (or several) variable(s) of interest, that are (partly) random,
and for which we would like to produce some predictions, within an environment
itself composed of random variables. Let us call our variable of interest Y, and the set
of the variables that compose the environment X. We may then follow the stylized
process:

- Step A: the researcher formulates a proposition of a model of the variable of
  interest, as a conditional expectation of its environment. The purpose of this step
  is to model the links between Y and X:

$$E[Y|X] = f(X). \tag{2.1}$$

This step, which may be very complex, can involve a discussion of the definition of
  Y and X, and the development of some sophisticated versions of $f()$. For example,
  it may include some dynamical representations if Y is a random process:

$$E\left[Y_t \mid \{X_s\}_{s\in[t-\tau:t]}\right] = f(X_t, X_{t-1}, X_{t-2}, \ldots). \tag{2.2}$$

*Here "t" is the current time index, "τ" is the farthest significant time-lag of X, and
"s" is the second time index defined between t−τ and t.*

- Step B: the statistical properties of the model are studied. This step may include a
  study of the distribution of the variable of interest Y and of the joint distribution of
  the environment variables X, but also a study of the distribution of the parameters
  of the model $f()$, in order to quantify their uncertainty, assess their robustness, etc.
- Step C: the model is used for a particular purpose. This may be the maximization
  of a given utility function, or the estimation of a risk measure, or any kind of goal
  that allows one to make a decision.
- Step D: the best models produced by steps A to C are selected. The variables they
  include, the functional forms they use, and the methods they employ are evaluated using performance and relevance measures, such as the p-value.

Some form of stress-testing of the models (for example using Monte-Carlo simulations) may be introduced either in step C as a validation of the model, or in step D as a selection criterion.

Since we model in order to act, our focus is entirely on the results obtained in step C.

Individual researchers often focus on step A to C, while D may be considered as a meta-problem which is related to how the literature evolves on a given topic. When performing a polymodel analysis, the steps are completed in a different manner. We first estimate a collection of univariate models, which corresponds to a repetition of step A. We then select the best models, which corresponds to step D, using criteria intentionally defined regarding the objective of step C. Step D itself is repeated, as we use it to consider the dynamic evolution of the collection of models through time. We primarily keep the information obtained in a multidimensional form, which allows us to derive a variety of indicators (e.g. the StressVaR (Coste et al. (2010)), or the Long-term Expectation (Guan, 2019)).

We perform step A repeatedly in an imprecise, simplified, and sub-optimal manner, but the multiplicity of models overcome this simplification since the collection of models is a very rich representation of the phenomenon we are studying. Let us take a toy example to clarify this point. We can model the returns of the S&P 500 by a collection of non-linear univariate models obtained from financial variables. Often, we assume the noise of the model to be Gaussian, which is of course quite simplistic, knowing that financial markets have fatter tails (see e.g. Platen and Rendek (2008) on this point). However, the non-linear modeling of oil, for example, may be able to partly capture the tail events of the S&P. And for a high number of independent variables in the factor set, it is likely that the non-linear modeling captures the tail events at some point (see Ye & Douady, (2019) for an example of market drawdown prediction using polymodels).

Questions about the robustness of the methods still need to be asked, but in a way that differs from the stylized research process presented above. Since we are entirely focused on achieving results in regard to step C, part of the work usually done in step B may become irrelevant. For example, we can observe that there is no particular reason for the transition from step B to C to be linear, it may even be highly non-linear in most of the cases. Following this reasoning, the quest for unbiased estimators becomes irrelevant, since they can't be used effectively to reach to final objective of step C. Hence, the robustness of the results obtained from the polymodel approach is often assessed through sensitivity analysis, or particular tests developed to measure the statistical significance of the results (see Chaps. 4, 5 and 6 of the present book). We thus do not discuss the problems addressed by step B, as although they are interesting by themselves, they are of secondary importance when considering step C as a central concern.

Note that apart from polymodels, another research field performs steps A and D directly regarding the objective of step C: machine learning (Friedman et al., (2001) provide a long introduction to the main techniques). From this perspective, polymodels could be considered as a machine learning technique.

Now that once the breadcrumb of the modeling approach of Polymodel Theory has been established, we can develop several points that will further clarify the concept. In order to do so, we organize the chapter as follows:

- As a final part of the introduction, we first review the current state of the literature on the topic of polymodels.
- We then formally define the notion of polymodel.
- This definition is followed by a discussion on how this object can be interpreted from an epistemological point of view.
- We then review the most salient advantages that are expected when using the technique, in econometrical terms. The use of a collection of univariate models is an alternative to the use of a multivariate model. We thus discuss these advantages with regard to standard alternatives, such as the classical linear regression estimated by OLS, Ridge regression, or Stepwise regression.
- We finally consider the challenges that Polymodel Theory raises, and conclude the chapter.

Some of the important questions that are raised when using polymodels in practice are:

- How do we estimate the univariate models?
- How do we select the variables?
- How do we aggregate the results?

The current chapter only proposes a theoretical overview of Polymodel Theory, and thus provides answers to the questions "what is a polymodel?" and "why using it?" but not to the question "how to use it?". The three practical questions listed above, which simply develop the more general question "how to use it?", may be approached in very different ways, that must be adapted to the empirical problem being tackled. Hence, presenting any of the techniques that we can use to answer these questions would cause the current chapter to lack generality. We thus restrict our explanations to the objective of presenting the notion of polymodels.

The literature on Polymodel Theory is still scarce. In its current form, there have been applications in finance, however the usage of collections of univariate linear models also exists outside of this field.

We first review the emerging financial literature on this topic. Apart from the initial paper of Coste et al. (2010), which introduces the notion, Polymodel Theory has already been used in a variety of applications in finance:

- Zhang (2019) built a clustering algorithm based on polymodel estimations, with applications to the equity market. The overall idea is that if two stocks react in the same manner to different factors, they are somehow similar. The clustering algorithm is used to design a statistical arbitrage trading strategy that delivers superior returns compared to the benchmark. The clustering algorithm is shown to outperform classic clustering methods (correlations, qualitative classification) in the context of statistical arbitrage.

- Ye and Douady (2019), and Kuang and Douady (2022) proposed some systemic risk indicators for equity indices based on polymodels. The indicators are essentially focused on the increase of the statistical significance of the links of a factor set with a stock index (Ye & Douady) and the concavity of the elementary models (Kuang & Douady).
- Guan (2019) used polymodels to produce some variations of traditional risk premia. He proved that the StressVaR, which is the risk indicator tested by Coste et al. (2010) for distinguishing risky hedge funds, is also a useful predictor of the cross-section of stock returns.

The literature on Polymodel Theory is thus still sparse, which justifies our proposal for a denser discussion of this framework of analysis.

Constructing a set of univariate models is quite an intuitive approach to modeling when the amount of data is too large to handle, a case in which a multivariate linear regression can bear some limitations. Indeed, having an uninvertible covariance matrix of predictors because there are more predictors than observations, or finding problems of multicollinearity, are concerns that are not confined to finance. Unsurprisingly, some traces and precedents of Polymodel Theory have been found in several disciplines:

- In genetics, the field of Genome-Wide Association Studies (GWAS) massively relied on linear versions of polymodels. GWAS encountered the problem of dealing with hundreds of thousands of predictors to predict a single target variable. Furthermore, there are more independent variables than observations. Classical regularization techniques have been used in an attempt to solve the problem of correlation among predictors, e.g. see de Vlaming and Groenen (2015) for Ridge regressions, Wu et al. (2009) for Lasso, or Liang and Kelemen (2008) for a literature review.
- In an analysis of driver fatality risk factors, Bose et al. (2013) used a set of univariate models to benchmark the coefficients obtained with a multivariate model.
- For the purpose of analysing epidemics, Bessell et al. (2010) began their study with a set of univariate models. They used the results of this polymodel to assess the statistical significance of the predictors, in order to select the most relevant of them to build a multivariate model.
- Ladyzhets (2019) proposed to analyse the probability space of a set of regression models to model financial time series. Although close to polymodels, as it represents the target variable using alternative models, the paper still uses multivariate models, thus losing some of the benefits of the former.

These last examples do not directly refer to Polymodel Theory as we present it in the current chapter, however, they show that the concerns we encounter when using multivariate regressions techniques are shared among several fields, making polymodels potentially interesting for mathematical applications outside of finance.

2.2 Mathematical Formulation

A polymodel can be defined as a collection of models, all equally valid and significant, that can be understood as a collection of relevant points of view on the same reality.

Mathematically, it can be equally formalized using Eq. (2.3) or Eq. (2.4):

$$\begin{cases} Y = \varphi_1(X_1) \\ Y = \varphi_2(X_2) \\ \quad \cdots \\ Y = \varphi_n(X_n) \end{cases} \tag{2.3}$$

$$\{Y = \varphi_i(X_i) \quad \forall i\}. \tag{2.4}$$

Here, Y is the target variable, X_i and φ_i are respectively the explanatory variable and the function of the ith model, with $i \in [1: n]$, and n the number of models (and factors).

The n models that we present here, called "elementary models", are models of a single variable. These models are all defined on the entire hyperspace of the explanatory variables $\mathbb{R}^n$. They do not interact with each other and they are all valid simultaneously.

The noise term ε_i is added to represent stochastic errors:[1]

$$\begin{cases} Y = \varphi_1(X_1) + \varepsilon_1 \\ Y = \varphi_2(X_2) + \varepsilon_2 \\ \quad \cdots \\ Y = \varphi_n(X_n) + \varepsilon_n. \end{cases} \tag{2.5}$$

Similarly, from Eq. (2.4):

$$\{Y = \varphi_i(X_i) + \varepsilon_i \quad \forall i\}. \tag{2.6}$$

2.3 Epistemological Foundations

Polymodel Theory can be considered from many different points of view. Even if we present applications clearly focused on financial mathematics, it is important to emphasize the philosophical roots that have led to the emergence of the theory.

---

[1] The noise term is often not of interest to us, but its representation is left to the discretion of the scientist, if needed.

2.3.1 A Statistical Perspectivism

The concept of perspectivism was first developed by the pre-Socratic philosopher Protagoras (c. 481 B.C.E.–c. 420 B.C.E.), whose thoughts we know through the dialogues of Plato (Lamb, 1967; Taylor & Lee, 2016). Perspectivism, the key to understanding Polymodel Theory, claims that we can't access a single and absolute truth. What we consider as true is often only true from the particular perspective adopted, and the formation of this perception of reality itself is dependent on the perspective in which it appears. It is an invitation to humbly understand that we are biased and have limited access and a limited understanding of the world.

One can easily understand the need to use several perspectives on the same question from the reflections of Pascal in his *Essay pour les coniques* (Clarke & Smith, 1928): "By the term conic section we mean the circumference of the circle, the ellipse, the hyperbola, the parabola and the rectilinear angle". All these perspectives on the same object are true and complementary, because none of them is able to fully describe the nature of the cone.

Thus, in the philosophical doctrines that flow from perspectivism, reality is the aggregation of all the perspectives that we have on it.

Similarly, the purpose of Polymodel Theory is to combine several descriptions of the same variable in order to get as close as possible to a full understanding of its nature. It provides a very rich description of reality, which is more than the sum of its parts, allowing us to understand very accurately some specific aspects of the considered variable. Hence, Polymodel Theory is a mathematical equivalent of philosophical perspectivism.

2.3.2 A Phenomenological Approach

The different elementary models that compose the polymodel are alternative descriptions of its variable of interest. But these descriptions are made in a particular way: they describe how the dependent variable *reacts* to each independent variable.

Phenomenology can be described as follows (Smith, 2018): "Literally, phenomenology is the study of 'phenomena': appearances of things, or things as they appear in our experience, or the ways we experience things, thus the meanings things have in our experience."

Thus, the point of interest of phenomenology is how the various phenomena that compose our experience interact with us. Polymodel Theory proposes a similar approach, by studying how various independent variables (i.e. an environment) interact with a variable of interest.

Such a position is by itself extremely meaningful, because it states that there is no interest given to the underlying mechanism of the dependent variable. A polymodel just describes how this variable behaves in various situations, that are as complete as the set of explanatory variables is. It does not explain why this behavior occurs, although it can help us to understand it. Hence, Polymodel Theory is closer to the physicist's approach to studying reality than to the economist's approach, as it primarily answers the question "how?" instead of the question "why?".

2.4 Comparison of Polymodels to Multivariate Models

2.4.1 Reducing Overfitting

Econometric models usually admit two components, a deterministic component, often called the mean equation, and a stochastic component, called the error term (see the introductions to econometrics by Stock and Watson (2015) and Seber and Lee (2012), or Alexopoulos (2010) for a concise paper about multivariate regression):

$$Y = f(X) + \varepsilon. \tag{2.7}$$

Here Y is the random variable of interest that we would like to explain, $f(X)$ is the mean equation composed of a set of random variables X and a function $f()$ that produces the association between Y and X, and ε is the error term, which follows some probability law usually centered at 0. When the set of random variables X contains only one variable, the model is called "univariate", while it is "multivariate" when it contains more than one variable. The division between deterministic ($f(X)$) and stochastic (ε) components is justified by the natural complexity of our world, in which there is usually a vast number of causes linked to the phenomenon that occurs, making each event partly unpredictable, simply because of the (current) inability of mathematical models to handle such a high level of complexity.

   This way of describing reality implies that only a part of the target variable's values can be described by the predictors, while the remaining part cannot and must be left unexplained. Assuming that there is an effective link between the target variable and the predictors, the aim of modeling in our context is to accurately represent this relation. This should be done by constructing a mean equation that represents the relation of the predictors to the deterministic part of the target variable, using only the information about the target variables that is contained in the predictors. This relation is formalized by a functional form, which is a stylized proxy for the true relation. The simplest functional form to manipulate is the linear one (Stock & Watson, 2015):

$$Y = X\beta + \varepsilon, \tag{2.8}$$

where β is a column vector of linear coefficients, and X is a matrix that contains a vector of ones when a constant is included (we assume this is always the case hereafter), and the vectors of the explanatory variables.

Given a particular sample of data, the task of the scientist is to choose a functional form and estimate its parameters to represent this relation. Of course, the goal being to explain the target variable, the importance of the deterministic part relative to the stochastic part[2] should be as high *as possible* in a good model.

However, the truly stochastic part of the model, i.e. the part of the target variable's values which is not related to the predictors, can always be spuriously modeled using the predictors. This is just a matter of using a sufficiently complex and appropriately parametrized function of the predictors. In-sample, this would greatly improve the usual measures of goodness of fit of the model, but in such a case, we would be modeling a link that does not exist since, by definition, there is no link between the predictors and the stochastic part of the target variable. Hence, after estimating the model using a particular sample of data, a direct consequence of this bad practice arises when new values of the predictors (outside of the original data set) are used to predict corresponding values of the target variable. The accuracy of the predictions is low, revealing the weakness of the initial model.

Observing a low predicting power out-of-sample of a model that exhibits a high level of goodness of fit in-sample is a typical definition of overfitting (Babyak, 2004).

Intuitively, one can see that it would be easy to find patterns in the data that seem to correspond to a link that does not exist, especially because we adapt these patterns to the target variable using the estimation of the functional form parameters.

This functional form, which characterizes the model, tends to adapt better to these spurious patterns when it is made more complex. This complexity depends on the number of parameters, and essentially comes from two sources: the complexity of the functional form associated to each variable, and the number of variables used in the model (Hawkins (2004) discusses these points in depth).

This is why increasing the number of predictors in a multivariate regression is well known to expose the danger of overfitting, so much so that econometricians on the subject like to joke by citing John von Neumann's famous quote: "With four parameters I can fit an elephant and with five I can make him wiggle his trunk". This is especially true when the number of observations is reduced, a situation which occurs frequently in real-life cases. Indeed, for a given number of observations, increasing the number of predictors in a multivariate model decreases the variance of the residuals, making the model 'better' in terms of goodness of fit, but only in appearance.

The problem becomes even worse when the number of predictors increases so much that it becomes higher than the number of observations. In such a case, the covariance matrix of the predictors $X'X$ is not invertible, leading to a failure of the usual ordinary least squares solution to the parameter's estimation of the linear model, defined as:

---

[2] In terms of variance explained.

$$\widehat{\beta} = [X'X]^{-1}X'Y. \tag{2.9}$$

The matrix inversion needed to compute the OLS estimator can be seen as a system of equations to solve, and in our case, there is no uniqueness of the solution. The problem is thus said to be "ill-posed" (Hadamard, 1902).

A very common approach to this problem is to regularize the regression using a penalty of the L^2-norm of the coefficient estimates (called Ridge regression (Hoerl & Kennard, 1988) or Tikhonov regularization). Recall that in ordinary least squares, the sum of the squared differences between the straight line and the data is minimized. The estimate of the parameter β, called $\widehat{\beta}$, is thus the solution of the optimization problem:

$$\min_{\beta} \ \|Y - X\beta\|_2^2. \tag{2.10}$$

In Ridge regression, the penalty of the L^2-norm is added as follows:

$$\min_{\beta} \ \|Y - X\beta\|_2^2 + \lambda\|\beta\|_2^2, \tag{2.11}$$

Here, "λ" is the Ridge penalty parameter.

Leading to the following shrunk estimates of the parameters:

$$\widehat{\beta}^{Ridge} = [X'X + \lambda I]^{-1}X'Y. \tag{2.12}$$

"I" is the $n \times n$ identity matrix.

On top of allowing the matrix inversion, the penalization of the L^2-norm automatically reduces the magnitude of the coefficients estimated by OLS, leading to better out-of-sample estimates (Van Dusen, 2016) proposes a comprehensive overview of this point). It is also possible to introduce in the optimization problem a penalty of both the L^1 and L^2-norm, a technique called Elastic Net regularization (Tibshirani, 1996, see also Zou and Hastie, 2005). While adding a penalization of the L^1-norm, some of the coefficients fall to zero, allowing for a selection of the independent variables. Ridge and Elastic Net regressions thus address the same problem as polymodels, which is an alternative when trying to prevent overfitting when modeling with a large number of variables.

However, in most of the applications of these approaches, the number of degrees of freedom of the model stays very low even after variable selection, which always raises some reservations about the resulting fit. This is a concern that is easily removed by using polymodels, because the number of degrees of freedom is greatly increased by the use of a single variable in each elementary model. Using a polymodel thus offers a simpler and more effective approach to solving these ill-posed problems.

To better understand how the multivariate models overfit compared to polymodels, we propose to illustrate our reasoning with a toy example. To tackle a familiar problem, we model the returns of the S&P 500 as a function of a set of n predictors (as presented in Eq. (2.7), in which Y is the S&P returns).

To test the spuriousness of the various approaches, we build a set of predictors that contains values randomly drawn from a Student's t-distribution with 4 degrees of freedom:[3]

$$X = 0 + \varepsilon$$

$$\varepsilon \sim t_{\nu=4}. \tag{2.13}$$

Hence, by construction, *there is no link between the target variable and the predictors* in our experiment. There is only noise to fit in the explanatory variables. To obtain $f()$, we compare three different modeling techniques:

- The multivariate linear model described by Eq. (2.8), estimated using the OLS estimator described by Eq. (2.9).
- The same multivariate model estimated using the Ridge estimator described by Eq. (2.12). The parameter λ, which is crucial for the estimation, is chosen using a 5-fold cross-validation (see Golub et al., 1979) on the use of cross-validation to choose the Ridge parameter).
- A polymodel, as described in Eq. (2.5), with elementary models defined as linear univariate models estimated with OLS.

This simple setting does not include non-linearity or variable selection, so that the results are only driven by the multivariate/univariate distinction. We estimate the parameters of each of these three models using the weekly returns of the S&P500 for the period from 2001-01-01 to 2003-12-31, which correspond to 157 observations. We then generate new values for X and use them with the previously estimated parameters to model the S&P's returns for the period from 2004-01-01 to 2006-12-31. Of course, we don't expect any of these predictions to be good, the experiment aims to reflect the ability of the different models to be trapped by fitting pure noise.

Thereby, we measure the overfitting by comparing the goodness of fit in-sample (the 2001–2003 period) to the goodness of fit out-of-sample (the 2004–2006 period). The goodness of fit is appreciated using the R^2, which is one minus the sum of squared residuals over the total sum of squares:

---

[3] Stock returns have been found to be quite well modeled by a Student's t-distribution with 4 degrees of freedom (Platen & Rendek, 2008).

$$R^2 = 1 - \frac{\sum\limits_{s=1}^{t} (y_s - f_t(x_s))^2}{\sum\limits_{s=1}^{t} (y_s - \bar{y}_t)^2}. \qquad (2.14)$$

Here, "t" is the current time index (i.e. end of 2003), "s" is the rolling time index in the window (i.e. it takes weekly date values between 01-2001 and 12-2003), "$\bar{y}_t$" is the average of the target variable between s= 1 and s=t, and "f_t" is the fitted function obtained with data available at date t.

More precisely, to assess overfitting, we measure the "Out-of-Sample R^2", which is the same formula as the "In-sample R^2" above, but with in-sample coefficients and out-of-sample data:

$$R^2_{oos} = 1 - \frac{\sum\limits_{s=t+1}^{t+157} (y_s - f_t(x_s))^2}{\sum\limits_{s=t+1}^{t+157} (y_s - \bar{y}_{t+157})^2}. \qquad (2.15)$$

The formula above reflects that new data, between $t + 1$ and $t + 157$, is used to recompute the R^2, but that the fitted function, $f_t()$, stays unchanged.

We then measure the *Spread* between the "In-Sample R^2" and the "Out-of-Sample R^2":

$$Spread = R^2_{IS} - R^2_{oos}. \qquad (2.16)$$

For the polymodel, we do not represent the spread of each elementary model but choose to present the average[4] spread among elementary models, for the sake of brevity.

Next we observe how the spread behaves as a function of the number of predictors n. We make this number vary between 1 and 156 (recall that there are 157 observations). This overfitting measure as a function of the number of predictors is compared for the three modeling techniques in Fig. 2.1:

The graph above represents the spread between the in-sample and out-of-sample goodness of fit, as a function of the number of regressors (for the polymodels the value displayed is the (unweighted, un-selected) average of the spreads among the different elementary models). For the multivariate models, the larger the number of independent variables, the larger the spread, thus the stronger the overfitting.

The spread of the multivariate OLS is so explosive when the number of regressors reaches the number of data points that it breaks the scale of the graph. Zooming makes it understandable (see Fig. 2.2):

---

[4]The average is a classical average, unweighted and un-selected. In practice, the usage of polymodels will include some form of selection and weighting of the elementary models.

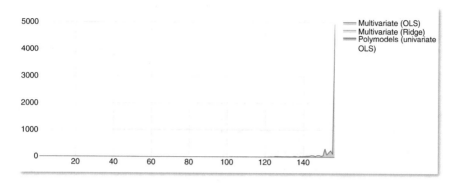

Fig. 2.1 Spreads between In-Sample vs Out-of-Sample R^2 as a function of the number of explanatory variables included in the regression

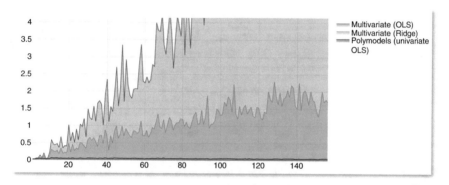

Fig. 2.2 Spreads between In-Sample vs Out-of-Sample R^2 as a function of the number of explanatory variables included in the regression (zoom)

The Ridge estimator behaves as expected in this toy example, notably reducing the overfitting compared to OLS in the multivariate case. Indeed, for the extreme case where $n = 156$, the spread of the Ridge estimator is roughly 3,000 times lower than that of the classical OLS estimator. However, even with regularization, it is clear that the overfitting grows as a function of the number of variables included in the multivariate models.

This is not true in the case of the average spread in the polymodel, which is (of course) asymptotically constant when the number of regressors n increases. For $n = 156$, the spread of the polymodel is 10^5 times lower than OLS and 34 times lower than Ridge in this particular case (Fig. 2.3).

This straightforward experiment also helps us to understand the nature of overfitting by comparing the *In-Sample* R^2 and *Out-of-Sample* R^2 of the three models:

In these three graphs, we have shown the *In-Sample* R^2 and the *Out-of-Sample* R^2 as a function of the number of independent variables, for the three fitting methods. In

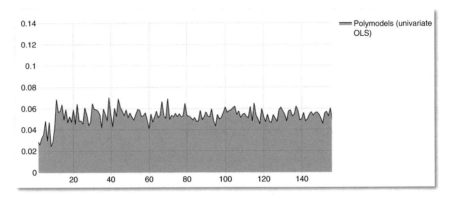

Fig. 2.3 Average spread between In-Sample vs Out-of-Sample R^2 as a function of the number of explanatory variables included in the regression (polymodels only)

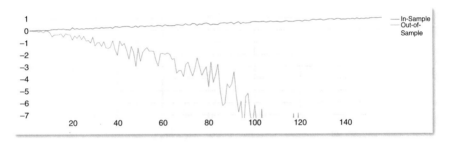

Fig. 2.4 Multivariate OLS: In-Sample vs Out-of-Sample R^2 as a function of the number of explanatory variables included in the regression

both of the multivariate models, we see that the *Spread* is not growing just because of the increase of the *In-Sample R^2*. The *Out-of-Sample R^2* worsens sharply, so dramatically that it quickly becomes the most important figure in the Spread computation. This illustrates the fact that overfitting not only results in too high expectations about the out-of-sample performance, but to a significant extent it actually also contributes to its deterioration, as we act only to follow past noise in that case. Note that the scale of the R^2 is quite different among the figures (Figs. 2.4, 2.5 and 2.6).

Of course, this toy example is a bit simplistic, far from the techniques we actually use,[5] but it allows us to explain the phenomenon of overfitting while giving a taste of the magnitude of its effects. Apart from the achieved levels of overfitting reduction, which are clearly dominated by polymodels, we may also question the use of a cross-

---

[5] Standard polymodel methods, as detailed by Zhang (2019), include non-linear modeling, shrinkage of the functional form, and selection of predictors.

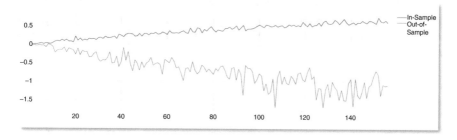

Fig. 2.5 Multivariate Ridge: In-Sample vs Out-of-Sample R^2 as a function of the number of explanatory variables included in the regression

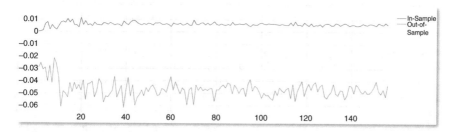

Fig. 2.6 Polymodels (Univariate OLS): In-Sample vs Out-of-Sample R^2 as a function of the number of explanatory variables included in the regression

validated regularized estimator when it seems that a repetition of simple OLS estimates performs better. We let the supporters of Occam's razor make their choice.

Another technique commonly used to handle a large number of potential candidate variables is stepwise regression. Stepwise regression builds the final model by starting with all variables in the model, and then eliminates some of them ('backward elimination'), or it starts with no variable, and then progressively introduces them into the model ('forward selection') (Hocking, 1976). The selection of the variables is based on a particular goodness of fit criterion, which often encourages a reduced number of variables in the model, such as adjusted R^2, the Akaike Information Criterion or Bayesian Information Criterion. A step-by-step procedure is then followed, in which the interest of adding/removing a variable of the model is evaluated using the goodness of fit measure at each step. Such a method may preserve a reasonable number of degrees of freedom (especially in the forward selection procedure), and also allows one to integrate non-linear variations of the independent variables more easily than Ridge regressions. However, such a repeated procedure creates a bias in the statistics used to assess its goodness of fit (Wilkinson & Dallal, 1981), and generally, the step-by-step procedure leads to overfitting (Flom & Cassell, 2007). Polymodels may be estimated in very different manners, hence one can easily avoid the trap of the step-by-step procedure.

Considering the usual alternatives to stepwise selection or regularization, the polymodel approach presents an effective way to reduce overfitting, since it cannot really be avoided in multivariate models when the number of potential independent variables is high.

2.4.2 Increasing Precision

Another concern of econometrics, though less often emphasized, is underfitting. The simplest way to model a relation between two variables is a linear model, however, there is no particular reason for the true reaction function to be a straight line. Even if the linear model delivers acceptable results for most of the cases, these results can be improved by introducing a smooth curve in place of the usual straight line. Indeed, if the true underlying model is non-linear, then a non-linear model should perform better than a linear one, thanks to its flexibility. However, the reverse is not true, if the underlying model is linear, then both linear and non-linear models have the ability to fit the straight line.[6] Moreover, non-linearity as a broad phenomenon has been found to be present in various financial situations, such as multi-factor modeling of equity markets (Caginalp & DeSantis, 2011, 2019), multi-factor modeling of interest rate volatility (Boudoukh et al., 1999), or lead-lag modeling of foreign exchange rates (Serletis et al., 2012).

In many domains, such as finance, a relevant increase in prediction accuracy can be decisive in terms of economic consequences, but the benefits that one can obtain from using non-linear modeling as a standard do not necessarily restrict to an increase in precision with respect to the linear model. In numerous situations, linear modeling leads to wrong modeling. Even in the cases where a linear model delivers good results *on average*, the events that occur in the tails of the distribution of the independent variables may demonstrate a very different response function with the target variable. This could have the worst consequences for the practitioner, since a model that is assumed to be good, and that seems to deliver consistently good results, may suddenly fail to make any accurate predictions. From this point of view, non-linear modeling is not just a viable method if one wants to slightly increase the accuracy of predictions, but it is a requirement to avoid surprises in extreme times.

A good example to illustrate this point is the perspective of fund of funds managers, who can be concerned with the multi-factor modeling of hedge fund returns. Hedge funds bear a significant tail risk, so their concave payoffs cannot be properly described with a linear factor model (Agarwal & Naik, 2004). A linear multi-factor model may be able to explain "normal times", non-extreme returns of hedge funds, but since the relation with their factor exposures is non-linear, it would

---

[6]Since non-linear models may be easily designed to fit a variety of functional forms, including linear ones.

be incapable of modeling their extreme returns, justifying the development of non-linear methods for this purpose (e.g. Cherny et al., 2010).

The use of non-linear functions in elementary models, for example polynomials, if fitted properly, allows one to increase the precision of the predictions, and to model extreme event[7] For example, we may define the function $\varphi_i(X_i)$ of Eq. (2.6) as:

$$\varphi_i(X_i) = \sum_{h=0}^{u} \beta_{i,h} X_i^h. \tag{2.17}$$

The function used above for the elementary model is a weighted sum of polynomials. The weights of such a model may be estimated with OLS. However, simple polynomials are highly correlated, and one of the assumptions of the OLS estimator is the linear independence of the regressors. Still, this assumption can be satisfied by orthogonalizing the polynomials. Assuming that the joint distribution of the factors is a Gaussian copula, we are led to use Hermite polynomials (Cherny et al., 2010). Chebyshev polynomials are known to be suitable for OLS estimations in the interval $[-1, 1]$ (Mason & Handscomb, 2002), an interval in which most of the financial returns are included. Finally, Guan (2019) proposed to numerically self-orthogonalize the polynomials. All these alternative choices have pros and cons, depending on the application. Note that plenty of possibilities exist for non-linearly approximating the functions $\varphi_i(X_i)$. For example, within the non-parametric world, the Nadaraya–Watson estimator (Nadaraya, 1964; Watson, 1964), also known as kernel regression, may deliver suitable results. Still, polynomial regression has the advantage of being extremely easy to implement, and thus may be quickly computed in real world situations.

Of course, our reasoning about the superiority of non-linear over linear techniques is conditioned by the fact that the polynomial model, if used, is not blindly fitted. Indeed, because the polynomial model is more complex, it is more likely to result in overfitting. To tackle this question, Zhang (2019) proposed to use regularization to shrink the parameters of polynomial elementary models.

The use of non-linear modeling allows for a very flexible functional form that, if used properly, is more data-driven than model-driven. This spirit of being reasonably adaptive to the data is also reflected in the variable assumptions made in the polymodels. Indeed, as we would not wish to miss a relevant perspective on reality to understand it, Polymodel Theory is an invitation to include any potentially important variable in the polymodel. There is virtually no limit to the number of variables that can (and should) be included in a polymodel *before the estimation procedure*. This exciting feature comes with the duty of systematically including a *selection method* in the estimation procedure of the polymodel, so that only the truly relevant variables are kept in the final estimated polymodel. At this stage, there is no standard for the selection method, which is therefore adapted to the various empirical

---

[7] Note that using non-linearity in polymodels is especially convenient because of the presence of a single variable per elementary model.

examples (Chaps 4 and 6 propose some solutions on this point). Thus, on top of using weak functional assumptions, Polymodel Theory also uses weak variable assumptions, which constitutes a good way to avoid model rigidity, driven by assumptions that may turn out to be false (from a general point of view, assuming a linear model can be seen as a strong assumption on the functional form of the model).

This extremely high level of flexibility of polymodels requires a careful estimation method. As stated, this estimation method should include a way to balance and/or select the different elementary models, but more importantly, the fit should not be too adaptive, otherwise overfitting can emerge again. Satisfying this need for a reasonable algorithm to fit the elementary models is the subject of Chap. 3.

The overall increase in precision which is introduced by non-linear modeling could be achieved using a stepwise regression. Such a method would also allow for automatic variable selection, however we already saw that the step-by-step procedure would lead to overfitting. In the case of Ridge and Elastic Net regularizations, introducing non-linear versions of predictors may overcome the lack of accuracy of a linear modeling, but it would clearly be at the cost of a greater risk of overfitting, since the number of coefficients in the model would necessarily increase. Hence, on the particular question of accurately fitting the patterns that are really present in large amounts of data, Polymodel Theory shows enviable characteristics compared to the standard alternatives.

2.4.3   Increasing Robustness

One of the most obvious benefits of using a polymodel, compared to a multivariate model, is the disappearance of multicollinearity. Multicollinearity, which we can expect to often appear often in large data samples, provides a non-robust estimate of the parameters, in the sense that the regression coefficients are highly sensitive to a small change in the data used for the estimation (see e.g. Belsley, 2014). This may be particularly problematic in the case of repeated fitting using a temporal rolling window, a technique often used in finance. Also, multicollinearity can make it difficult to reliably identify the independent variables that are effectively linked with the target variable that we are trying to explain. As it increases the standard errors of the affected coefficients, it spuriously reduces the p-value of the parameter if computed via the usual t-statistic. While the problem of multicollinearity among predictors can be undertaken using Ridge regression at the cost of accuracy (Hoerl & Kennard, 1970), in the framework of Polymodel Theory multicollinearity just doesn't appear, which makes it more suitable for obtaining a robust estimation of the reaction function of each predictor.

We propose to draw from the real world another toy example to illustrate this particular point. Again, we consider the problem of fitting a multi-factor model with a multivariate OLS, a multivariate Ridge and a simple linear polymodel. The target variable is still the weekly returns of the S&P 500, and we continue to study it over

Table 2.1 Sensitivity of estimates to changes in the set of explanatory variables

	Multivariate (OLS)	Multivariate (Ridge)	Polymodels (Univariate OLS)
Count	1,806	1,806	1,806
Mean	76%	28%	0%
Min	0%	0%	0%
Quantile 25%	2%	1%	0%
Quantile 50%	10%	5%	0%
Quantile 75%	39%	18%	0%
Max	9.110%	1.556%	0%
Std	4.12	0.93	0.00
Kurtosis	233.30	88.70	0.00
Skewness	13.83	8.19	0.00

the period from 2001-01-01 to 2003-12-31 (157 observations). However, the set of predictors we use to predict it is no longer composed of random draws. Instead, we use 43 US equity indices, most of which are sectorial indices.[8] We first estimate the three different models using the 43 explanatory variables and collect their associated parameters. Next, we re-estimate the three models while removing one of the predictors, thus keeping 42 of them. If the initial estimates were robust, they should be stable, and thus be exactly the same as the coefficients estimated with 43 variables in the multi-factor model. For each model, we measure the absolute change in coefficients, expected to be equal to 0, with the following basic metric:

$$\Delta_{coeff} = \left| \frac{\widehat{\beta}_{42} - \widehat{\beta}_{43}}{\widehat{\beta}_{43}} \right|. \tag{2.18}$$

This procedure is then repeated 43 times, each time changing the variable that is removed from the predictor set, so that we can collect $42 \times 43 = 1,806$ values for Δ_{coeff} for the three models. Here are the descriptive statistics of these values (Table 2.1):

The average absolute change in the coefficient is 76% for the multivariate OLS, 28% for the multivariate Ridge, and obviously 0% for the polymodel. These extremely large changes in the multivariate case result from a high level of multicollinearity in the predictors. Indeed, their average correlation is 56%.

---

[8] For the purpose of replication, here is the complete list of Bloomberg tickers: CBNK Index, CCMP Index, CIND Index, CINS Index, CUTL Index, CXBT Index, DJUSBM Index, DJUSCY Index, DJUSEN Index, DJUSFN Index, DJUSHC Index, DJUSIN Index, DJUSNC Index, DJUSRE Index, DJUSUTT Index, DWCTEC Index, INDU Index, IXK Index, R1HLTH Index, R2COND Index, R2CONS Index, R2HLTH Index, R2UTIL Index, RGUSDS Index, RGUSES Index, RGUSHS Index, RGUSMS Index, RGUSSS Index, RGUSTS Index, RGUSUS Index, RIY Index, RTY Index, S5COND Index, S5CONS Index, S5ENRS Index, S5FINL Index, S5HLTH Index, S5HOTR Index, S5INDU Index, S5INFT Index, S5MATR Index, S5TELS Index, S5UTIL Index.

Of course, there are plenty of cases where the predictors used are only lowly correlated, or not correlated at all. Still, the higher the number of predictors used in the model, the more likely the appearance of multicollinearity for some of the explanatory variables. The simple example we presented shows that even if the Ridge estimator is indeed effective at reducing parameter instability, its results remain imperfect, while any multicollinearity totally disappears with polymodels, which thus provide a radical tool to solve this problem.

In order to guarantee that the estimated model is reliable, one of the standard hypotheses in modeling is that the variance of the residuals is constant for all observations of the sample used for the fit. Homoskedasticity may be difficult to ensure while working with time series, which can show clusters of volatility (in finance, this phenomenon has been notably observed by Mandelbrot, 1997; Engle, 1982). Hence, the longer the window used for the estimation, the less likely the homoskedasticity assumption is fulfilled. When this condition is not fulfilled, the standard errors of the coefficients estimated by OLS are biased, leading again to spurious levels of p-values based on t-statistics, and thus potentially to a wrong identification of the most relevant variables to consider in the analysis. Note that when trying to avoid overfitting, multivariate models must rely on the largest temporal depth available. In contrast, polymodels allow one to reduce the temporal depth of the estimations while keeping a satisfying number of degrees of freedom, which is less prone to overfitting. Polymodels are thus more likely to satisfy the hypothesis of constant variance in the sample while trying to simultaneously prevent overfitting. Hence, in addition to avoiding multicollinearity, the robustness of the modeling is also better using polymodels instead of multivariate models from the particular perspective of respecting the homoskedasticity assumption.

When modeling with several variables, the practitioner frequently encounters the problem of missing data. Missing observations in the matrix of the predictors prevents estimation, thus it leads either to drop all the observations of the other variables that have the same index (e.g. simply removing a date), or to fully drop the entire variable itself. When using a large number of variables, this question becomes more and more problematic, since a wide range of observations often miss, and these observations may not miss simultaneously. Whatever the choice taken, the use of any multivariate regression technique leads to the removal of an important part of the available data. The robustness of the model is also linked to the number of observations used to estimate its parameters. Again, polymodels allow to overcome this concern. Since each variable is fitted independently using its own elementary model, there is no need to remove any observations, the number of observations used in each elementary model can be different. Naturally, this difference should then be taken into account in the tests used to assess the statistical significance of each variable. By keeping all the observations, polymodels thus increase the robustness of the estimations compared to multivariate models.

Multivariate modeling techniques can only address these concerns about the robustness of the estimations with difficulty. Although Ridge regressions may partly solve the problem of multicollinearity, none of the multivariate techniques are able to respect the raw structure of the data (dynamics of variance, missing observations) as well as polymodels do. These findings position the polymodel technique as a suitable one to work with large amounts of data.

2.5   Considerations Raised by Polymodels

2.5.1   Aggregation of Predictions

One of the usual questions raised when Polymodel Theory is presented is which method to use to aggregate the predictions.

First, it is important to note that Polymodel Theory offers more than an aggregated prediction. Since its purpose is to provide a quasi-exhaustive representation of the links that a variable maintains with its whole environment, there is a lot to learn from this representation itself. The researcher can focus on the dynamics of the strength of these links, their amount of non-linearity, etc. Also, the elementary models may be used to produce measures other than a prediction of the target variable, such as the Value at Risk. Thus, very different measures representing different aspects of the system may be produced.

The aggregation of the measures that investigate the different dimensions of the polymodel can be done in very different ways. We may consider the entire distribution of these measures, as well as maxima, minima or extreme quantiles that depict the reactions of the system in stressed conditions. The volatility over time of the measures brings information about their instability. It is also possible to consider the ratio of the measure to its historical mean, in order to get a meaningful value of its relative present level.

Finally, aggregating the predictions of the different elementary models together raises the question of the correlations of different predictors.

In order to effectively understand the stakes of this question, we approach it through the metaphor of an amphitheater.

Let us consider an amphitheater filled by students from the department of finance of a university. We would like to use their knowledge to predict the future returns of a financial asset. Before joining the amphitheater, the students have been selected to be the most skilled in the university for this particular task. The selection has been conducted using a multiple-choice questionnaire. Thus, a small proportion of the bad students, who have answered the questionnaire following a random guessing procedure, may have been selected by chance. The test used to determine the best students is fallible, as all tests are.

These spuriously selected students will not have the most common profile among those we may encounter in the amphitheater. Most of the students selected are skilled, however, since they all come from the same university, they share the same knowledge and ideas about financial markets.

However, some of the students started their studies in other universities, even in different fields in a few cases, and some of them study more than others, looking for complementary information outside of the lectures, hence, some original views may be expressed by those students.

The predictions made by the students are anonymous, thus we can't know who these original students are. We only observe the predictions, which are repeated for several rounds, trying each time to predict the return of the financial asset for the next period.

It is quite transparent that the amphitheater is a polymodel, and that each of the students represents an elementary model. Also, we clearly understand that the predictions of most of the students, even if relevant, would be correlated, since the way they represent the market would be essentially the same. The question is then how to distinguish the wise, the fool and the crowd. This question can be addressed in several ways, and we propose to tackle it through a measure that considers both originality and believability, which is presented in Chap. 6.

2.5.2 Number of Variables Per Elementary Model

The choice of including only a single variable per elementary model needs to be discussed. Indeed, most of the benefits of the polymodels are preserved if the elementary models are composed of several variables. Such a construction is also in line with the epistemological approach of Polymodel Theory.

Nonetheless, including several variables in elementary models would not allow us to have a high level of precision (i.e. non-linear modeling) without overfitting. Some of the other benefits, for example the absence of multicollinearity, would be lost.

Apart from statistical considerations, the use of univariate elementary models simplifies a lot the analysis, since it is clear that the metrics that we measure on the elementary models are associated with a single, well-identified factor.

Thus, univariate elementary models seem to be the most appropriate choice for the construction of the polymodels.

2.6 Conclusions

Polymodel Theory is an intuitive approach that has been used in different fields for a long time, but it has often been hastily rejected in favor of multivariate modeling. However, the first applications in finance show that the method provides a rich framework, particularly favorable to the non-linear modeling of big data, with a large panel of applications outside of the scope of multivariate modeling.

Formalizing a tool that is increasingly used in the recent literature, we have shown that the multi-univariate approach of polymodels has many favorable qualities from an econometrical point of view. In particular, these advantages are in line with the current stakes of finance, in which techniques that can handle large amounts of data in a data-driven, robust, accurate, and non-overfitted manner are required. From this perspective, Polymodel Theory can be seen as a machine learning method.

Polymodel Theory also has philosophical benefits: from the point of view of epistemology it seems to be a more reasonable approach than multivariate modeling, since it provides several representations of the same object (that can be weighted according to believability) in lieu of a single, immutable representation.

The non-linearity of polymodels makes it possible to efficiently tackle regime changes in the market dynamics that are "spatial", that is, due to the size of the moves, rather than to a random "temporal" event. This is in line with, for instance, hidden Markov models in which the "beta" of stocks with respect to certain indices depends on the regime.

The challenges of estimating, selecting and aggregating the predictions are important topics that are addressed in the following chapters. Managing these questions properly is the cornerstone of the use of polymodels. Still, if properly handled, it makes Polymodel Theory a powerful modeling method, and to some extent, a superior alternative to most of the traditional multivariate regression techniques.

References

Agarwal, V., & Naik, N. Y. (2004). Risks and portfolio decisions involving hedge funds. *The Review of Financial Studies, 17*(1), 63–98.

Alexopoulos, E. C. (2010). Introduction to multivariate regression analysis. *Hippokratia, 14*(Suppl 1), 23.

Aris, R. (1994). *Mathematical modelling techniques*. Courier Corporation.

Babyak, M. A. (2004). What you see may not be what you get: a brief, nontechnical introduction to overfitting in regression-type models. *Psychosomatic Medicine, 66*(3), 411–421.

Belsley, D. A. (2014). *Conditioning diagnostics*. Wiley.

Bessell, P. R., Shaw, D. J., Savill, N. J., & Woolhouse, M. E. (2010). Estimating risk factors for farm-level transmission of disease: foot and mouth disease during the 2001 epidemic in Great Britain. *Epidemics, 2*(3), 109–115.

Bose, D., Arregui-Dalmases, C., Sanchez-Molina, D., Velazquez-Ameijide, J., & Crandall, J. (2013). Increased risk of driver fatality due to unrestrained rear-seat passengers in severe frontal crashes. *Accident Analysis & Prevention, 53*, 100–104.

Boudoukh, J., Richardson, M., Stanton, R., & Whitelaw, R. F. (1999). *A multifactor, nonlinear, continuous-time model of interest rate volatility* (No. w7213). National Bureau of Economic Research.

Caginalp, G., & DeSantis, M. (2011). Nonlinearity in the dynamics of financial markets. *Nonlinear Analysis: Real World Applications, 12*(2), 1140–1151.

Caginalp, G., & DeSantis, M. (2019). Nonlinear price dynamics of S&P 100 stocks. *Physica A: Statistical Mechanics and its Applications, 547*(3), 122067.

Cherny, A., Douady, R., & Molchanov, S. (2010). On measuring nonlinear risk with scarce observations. *Finance and Stochastics, 14*(3), 375–395.

Clarke, F. M., & Smith, D. E. (1928). Essay Pour Les Coniques' of Blaise Pascal. *Isis, 10*(1), 16–20. www.jstor.org/stable/224736

Coste, C., Douady, R., & Zovko, I. I. (2010). The StressVaR: A new risk concept for extreme risk and fund allocation. *The Journal of Alternative Investments, 13*(3), 10–23.

Davis, P., Hersh, R., & Marchisotto, E. A. (2011). *The mathematical experience*. Springer Science & Business Media.

De Vlaming, R., & Groenen, P. J. (2015). The current and future use of ridge regression for prediction in quantitative genetics. *BioMed Research International, 2015*, 143712.

Engle, R. F. (1982). Autoregressive conditional heteroscedasticity with estimates of the variance of United Kingdom inflation. *Econometrica: Journal of the Econometric Society, 50*, 987–1007.

Flom, P. L., & Cassell, D. L. (2007, November). Stopping stepwise: Why stepwise and similar selection methods are bad, and what you should use. In *NorthEast SAS Users Group Inc 20th Annual Conference* (pp. 11–14).

Friedman, J., Hastie, T., & Tibshirani, R. (2001). *The elements of statistical learning* (Vol. 1, No. 10). Springer Series in Statistics.

Golub, G. H., Heath, M., & Wahba, G. (1979). Generalized cross-validation as a method for choosing a good ridge parameter. *Technometrics, 21(2)*, 215–223.

Guan, Y. (2019). *Polymodel: application in risk assessment and portfolio construction* (Doctoral dissertation, State University of New York at Stony Brook).

Hadamard, J. (1902). Sur les problèmes aux dérivées partielles et leur signification physique. *Princeton University Bulletin, 13*, 49–52.

Hawkins, D. M. (2004). The problem of overfitting. *Journal of Chemical Information and Computer Sciences, 44(1)*, 1–12.

Hocking, R. R. (1976). A Biometrics invited paper. The analysis and selection of variables in linear regression. *Biometrics, 32(1)*, 1–49.

Hoerl, A., & Kennard, R. (1988). Ridge regression. In *Encyclopedia of statistical sciences* (Vol. 8). Wiley.

Hoerl, A. E., & Kennard, R. W. (1970). Ridge regression: Biased estimation for nonorthogonal problems. *Technometrics, 12(1)*, 55–67.

Kuang, Y., & Douady, R. (2022). Crisis risk prediction with concavity from Polymodel. *Journal of Dynamics & Games, 9(1)*, 97. https://doi.org/10.3934/jdg.2021027

Ladyzhets, V. (2019). Probability space of regression models and its applications to financial time series. *Model Assisted Statistics and Applications, 14(4)*, 297–310.

Liang, Y., & Kelemen, A. (2008). Statistical advances and challenges for analyzing correlated high dimensional SNP data in genomic study for complex diseases. *Statistics Surveys, 2*, 43–60.

Mandelbrot, B. B. (1997). *The variation of certain speculative prices. In Fractals and scaling in finance* (pp. 371–418). Springer.

Mason, J. C., & Handscomb, D. C. (2002). *Chebyshev polynomials*. CRC Press.

Nadaraya, E. A. (1964). On Estimating Regression. *Theory of Probability and its Applications, 9(1)*, 141–142.

Platen, E., & Rendek, R. (2008). Empirical evidence on Student-t log-returns of diversified world stock indices. *Journal of Statistical Theory and Practice, 2(2)*, 233–251.

Plato. (1967). *Plato II: Laches, protagoras, meno, euthydemus* (W. R. M. Lamb, Trans.). Harvard University Press.

Seber, G. A., & Lee, A. J. (2012). *Linear regression analysis* (Vol. 329). John Wiley & Sons.

Serletis, A., Malliaris, A. G., Hinich, M. J., & Gogas, P. (2012). Episodic nonlinearity in leading global currencies. *Open Economies Review, 23(2)*, 337–357.

Smith, D. W. (2018). Phenomenology. In *The Stanford encyclopedia of philosophy* (Summer 2018 Edition). Edward N. Zalta (ed.). https://plato.stanford.edu/archives/sum2018/entries/phenomenology/

Stock, J. H., & Watson, M. W. (2015). *Introduction to econometrics* (3rd ed.). Pearson.

Taylor, C. C. W., & Lee, M.-K. (2016). The sophists. In *The Stanford encyclopedia of philosophy* (Winter 2016 Edition), Edward N. Zalta (ed.). https://plato.stanford.edu/archives/win2016/entries/sophists/.

Tibshirani, R. (1996). Regression shrinkage and selection via the lasso. *Journal of the Royal Statistical Society. Series B (Methodological), 58*, 267–288.

Van Dusen, C. (2016). Methods to prevent overwriting and solve ill-posed problems in statistics: Ridge Regression and LASSO. *Preprint submitted to Colorado College Department of Mathematics September 16.*

Watson, G. S. (1964). Smooth regression analysis. *Sankhya: The Indian Journal of Statistics, Series A., 26*, 356–372.

Wilkinson, L., & Dallal, G. E. (1981). Tests of significance in forward selection regression with an F-to-enter stopping rule. *Technometrics, 23*(4), 377–380.

Wu, T. T., Chen, Y. F., Hastie, T., Sobel, E., & Lange, K. (2009). Genome-wide association analysis by lasso penalized logistic regression. *Bioinformatics, 25*(6), 714–721.

Ye, X., & Douady, R. (2019). Systemic risk indicators based on nonlinear polymodel. *Journal of Risk and Financial Management, 12*(1), 2.

Zhang, J. (2019). *Statistical arbitrage based on stock clustering using nonlinear factor model* (Doctoral dissertation, State University of New York at Stony Brook).

Zou, H., & Hastie, T. (2005). Regularization and variable selection via the elastic net. *Journal of the Royal Statistical Society: Series B (Statistical Methodology), 67*(2), 301–320.

Chapter 3
Estimation Method: The Linear Non-Linear Mixed Model

Abstract We introduce the Linear Non-Linear Mixed (LNLM) model as an effective method to produce non-linear univariate models, with the primary concern of reducing overfitting. We show using numerical simulations that the LNLM model is able to successfully detect patterns in noisy data, with an accuracy similar to or better than data-driven modeling alternatives. We find that our algorithm is computationally efficient, an essential characteristic for machine learning applications often involving a large number of estimations.

Keywords Non-Linear modeling · Polynomial regression · Polymodel theory · Regularization · Overfitting · Data-driven · Non-parametric · Univariate regression

3.1 Introduction

As explained in Chap. 2, one of the most powerful advantages of a polymodel over a classical multi-factor model is the possibility of more accurately fitting the target variable, without restricting ourselves to simple linear functional forms. Plenty of alternatives are available in the domain of data-driven modeling, but for most of the techniques, the estimation process may be heavy in terms of computational resources, which is a particularly important point in the polymodels framework. Indeed, even simple applications of Polymodel Theory may involve millions to billions of fits. For example, one may want to fit a polymodel using a thousand factors updated daily for a simulation of 20 years, which directly leads to $1,000 * 252 * 20 = 5,040,000$ fits. It is easy to imagine that we may have several target variables and/or a higher frequency in our data. The computational time thus matters a lot in the big data framework that polymodels involve.

The simple solution that practitioners originally used to estimate an elementary model was to use a weighted sum of polynomials. Douady, along with Molchanov and Cherny (2010), introduced the use of Hermite polynomials, which exhibit interesting properties that limit the correlations among the polynomials when used together in a regression.

Recall that the Hermite polynomials are defined as:

$$H_h(x) \overset{\lrcorner}{=} (-1)^h e^{\frac{x^2}{2}} \frac{d^h}{dx^h} e^{\frac{-x^2}{2}}. \tag{3.1}$$

We then simply use the following non-linear model to estimate the elementary models:

$$\varphi_i(X_i) = \sum_{h=1}^{u} \beta_{h,i} H_h(X_i) + \varepsilon_i. \tag{3.2}$$

Here the betas are the estimates obtained from OLS (4 polynomial terms are usually enough to reach a sufficient level of accuracy in practice), i.e.:

$$\widehat{\beta}^{OLS} = [X'X]^{-1}X'Y. \tag{3.3}$$

The polynomial solution has many advantages, since OLS just requires matrix inversions, a task that can be done efficiently nowadays, and that can be easily parallelized. Indeed, parallel computing is one of the keys to an efficient use of Polymodel Theory by the practitioner. Also, polynomial combinations can capture in a smooth and data-driven functional form the underlying link between the independent and the target variables.

However, these advantages come at a cost: overfitting. Using the polynomial model leads one to re-introduce several artificial exogenous variables in the elementary models, and because of their non-linear nature, which is particularly adaptive, this creates a favorable ground for overfitting.

The Linear Non-Linear Mixed model comes as an answer to this concern. On one hand, a weighted sum of polynomials models the data so well that it fits some noise along with the underlying relation between variables. On the other hand, a linear model would provide a more robust fit, but at the very high cost of offering only a naïve and simplistic representation of reality. The LNLM model proposes to mix both models and only retain their advantages.

Such a regularized fit may of course be achieved using other techniques.

The polynomial model may be regularized using Ridge estimates to get the betas of the model, which simply consists in adding a penalization on the diagonal of the covariance matrix of the predictors:

$$\widehat{\beta}^{OLS} = [X'X]^{-1}X'Y \Rightarrow \widehat{\beta}^{Ridge} = [X'X + \lambda I]^{-1}X'Y. \tag{3.4}$$

A proper value for the parameter λ can be obtained using cross-validation (see Golub et al., 1979), a process which is achieved numerically, by testing different values of λ and their associated (pseudo-) out-of-sample goodness of fit. The problem with such a method is that it requires the penalized covariance matrix to be inverted each time a value of λ is tested. Hence, performing a 10-fold cross-validation while evaluating 10 different λ leads to 100 matrix inversions, which may be of low performance in terms of computational time.

Of course, other techniques may be appropriate to propose a data-driven fit which is less over-fitted than a polynomial model estimated with OLS. Among them, we retain the Nadaraya–Watson estimator (Nadaraya, 1964; Watson, 1964) as a standard benchmark that we use to compare the performance of the LNLM model.

The aim of this chapter is thus to present the LNLM model, and to show its interest in terms of improving out-of-sample goodness of fit as well as in terms of computational time. As for many other artificial intelligence techniques (see Chap. 2), our approach is focused on the effectiveness of the model when it is used (here to produce non-overfitted predictions), more than on a deep understanding of its statistical properties, which are consequently not analyzed in the present chapter.

We thus organize the chapter as follows:

- First, we present the LNLM model, through a formal definition, and we give an explanation of the fitting procedure.
- We then present a methodology designed to evaluate the efficiency of the model using a large panel of simulations.
- The results of the simulations are presented and discussed, and we eventually conclude the chapter.

3.2 Presentation of the LNLM Model

3.2.1 Definition

The LNLM model aims to represent a target variable as follows:

$$Y = LNLM(X). \tag{3.5}$$

This representation is done using the following definition:

$$LNLM(X) \overset{\Delta}{=} \bar{y} + \mu \sum_{h=1}^{u} \widehat{\beta}_h^{NonLin} H_h(X) + (1 - \mu)\widehat{\beta}^{Lin} X + \varepsilon. \tag{3.6}$$

Here $0 \leq \mu \leq 1$ is the parameter that allows one to control for the potential[1] non-linearity, and is thus called the non-linearity propensity parameter, and $\bar{y}$ is the mean of the target variable.

---

[1]The polynomial fit can result in a linear model.

The LNLM model thus simply consists of a regularization of a polynomial model[2] by a linear one. This point is very important, and creates a significant difference with a classical shrinkage of the parameters using a Ridge regression[3] (Hoerl & Kennard, 1988). Indeed a natural idea that would first come to mind to reduce the overfitting induced by the use of the polynomials would be to simply shrink the OLS estimates of the model, as done by Zhang (2019). Other recognized alternatives would be the LASSO (Tibshirani, 1996) or the Elastic Net (Zou & Hastie, 2005) approaches, however these two methods includes a penalty of the L^1-norm, which often leads one to discard some of the covariates. This makes a lot of sense in a parsimonious selection of different independent variables, but in our case, where we only use different variations of the same independent variable, we expect that keeping all the polynomial terms would result in a more balanced aggregated function.

The motivation for using the LNLM model instead of the Ridge model is primarily theoretical. Our prior belief is that the over-fitting of the polynomial model comes from the non-linear terms of the model only. In other words, we never expect the linear part of the model to over-fit the data. In Ridge regression, shrinking the linear term cannot be avoided. Recall that the estimates of the Ridge regression are defined as:

$$\widehat{\beta}^{Ridge} = [X'X + \lambda I]^{-1}X'Y, \quad s.t. \quad \lambda \geq 0 \tag{3.7}$$

Here λ is an integer, the shrinkage parameter, and "I" is the $u \times u$ identity matrix.

In the case of the polynomial model, the matrix $X'X$ would be 4×4, so we may replace the 4×4 identity matrix I by the matrix J [4]:

$$J = \begin{bmatrix} 0 & 0 & 0 & 0 \\ 0 & 1 & 0 & 0 \\ 0 & 0 & 1 & 0 \\ 0 & 0 & 0 & 1 \end{bmatrix} \tag{3.8}$$

This solution is tempting, since we may expect that the linear term will no longer be penalized using the matrix J. However, we still have to invert the matrix $[X'X + \lambda J]$, which is equivalent to solving a four equation system. In such a system, only the

---

[2] Similar results may be obtained with a different set of polynomials than the Hermite polynomials. Empirical tests have not found that this particular set of polynomials provide superior predictive power, thus they are just kept for convenience, legacy and as a tribute to the founding paper of Douady.

[3] More formally known as Tikhonov regularization.

[4] In Ridge regression, the target variable is centered before parameter estimation, and only polynomials of degree 1 to 4 are used. The constant of the polynomial model is the average of the target variable. This procedure is standard and avoids shrinking the constant term.

second, third and fourth equations are modified, and the first equation stays unchanged after the addition of the penalty term in the covariance matrix. But when solving the system, the solution of each equation depends on the solution of all others, thus the solution of the first equation of the system is also modified, and as a result, the linear coefficient is changed.

The LNLM model is thus a response to the limit of the Ridge regression in the very special case of the fitting of a univariate polynomial model (maintaining the assumption that the overfitting only comes from non-linearity).

3.2.2 Fitting Procedure

The structure of the LNLM model prevents a global direct fit using OLS, first because the result would just be a large overfitted polynomial model, secondly because the perfect correlation of the linear terms in the model would preclude matrix inversion.

Therefore, we design a three-step fitting procedure, where we first choose the value of μ, the non-linearity-propensity parameter, then separately estimate the non-linear model and the linear model and finally combine all the ingredients of the LNLM to get the final fit. Steps 2 and 3 are trivial, hence we focus on detailing step 1 in what follows.

The aim of our approach being to reduce overfitting, we structure our methodology around the concept of cross-validation. More precisely, we use a variation of k-fold cross-validation to numerically approach the value of μ that minimizes the overfitting, i.e. the out-of-sample error. The choice of cross-validation is motivated by its proven ability to reduce overfitting (Moore, 2001).

In order to do this, we first split the target variable data into k sub-samples, called "folds". Stratified K-Folds encompasses several techniques that aim to keep sub-samples representative of the global distribution of the target variable. If we have q observations available, we split the data into $\lceil q/k \rceil$ quantiles. Thus, each of the quantile buckets contains exactly k observations (except the last one if q/k is not an integer). If we take the example of a 5-fold cross-validation performed over 50 observations, we get a distribution split into 10 quantiles of 5 observations (Fig. 3.1):

In each complete quantile, we then randomly assign a fold identifier to each observation. All the fold identifiers are assigned with equal probability, and the same identifier cannot appear more than once in the same quantile bucket. The potentially incomplete quantile is the only one in which not all of the k identifiers may be represented. We finally just group the observations by fold identifier, and we get k folds, each containing observations from all the quantiles previously defined. Hence, all the folds contain data that is representative of the full sample distribution of the target variable. In our previous example, we get 5 folds, each of which include 10 observations drawn from the 10 quantiles of the initial distribution (Fig. 3.2):

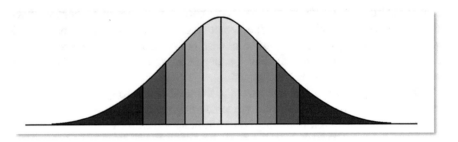

Fig. 3.1 Stylized representation of the quantile split for stratified cross-validation

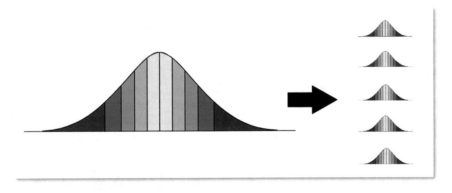

Fig. 3.2 Stylized representation of the folds distributions for stratified cross-validation

Once the splitting of the data into folds is performed, we fit the linear model and the non-linear model separately using the last $k-1$ folds. Then we use the first fold, which hasn't been used to estimate the OLS parameters, to numerically compute the value of μ that minimizes the root mean squared error (RMSE) inside this pseudo out-of-sample fold:

$$\min_{\mu_1} \sqrt{\frac{1}{q}\sum_{d=1}^{q}\left(y_{1,d} - \left[\bar{y}_{2,3,\ldots,k} + \mu_1\sum_{h=1}^{u}\widehat{\beta}_{2,3,\ldots,k,h}^{NonLin}H_h(x_{1,d}) + (1-\mu_1)\widehat{\beta}_{2,3,\ldots,k}^{Lin}x_{d,1}\right]\right)^2}$$

(3.9)

This numerical choice is repeated k times, each time changing the pseudo out-of-sample fold that is used to determine μ:

$$\begin{cases}
min_{\mu_1}\sqrt{\frac{1}{q}\sum_{d=1}^{q}\left(y_{1,d}-\left[\bar{y}_{2,3,\dots,k}+\mu_1\sum_{h=1}^{u}\widehat{\beta}_{2,3,\dots,k,h}^{NonLin}H_h(x_{1,d})+(1-\mu_1)\widehat{\beta}_{2,3,\dots,k}^{Lin}x_{1,d}\right]\right)^2} \\[2mm]
min_{\mu_2}\sqrt{\frac{1}{q}\sum_{d=1}^{q}\left(y_{2,d}-\left[\bar{y}_{1,3,\dots,k}+\mu_2\sum_{h=1}^{u}\widehat{\beta}_{1,3,\dots,k,h}^{NonLin}H_h(x_{2,d})+(1-\mu_2)\widehat{\beta}_{1,3,\dots,k}^{Lin}x_{2,d}\right]\right)^2} \\[2mm]
\dots \\
\dots \\
\dots \\
min_{\mu_k}\sqrt{\frac{1}{q}\sum_{d=1}^{q}\left(y_{k,d}-\left[\bar{y}_{1,2,\dots,k-1}+\mu_k\sum_{h=1}^{u}\widehat{\beta}_{1,2,\dots,k-1,h}^{NonLin}H_h(x_{k,d})+(1-\mu_k)\widehat{\beta}_{1,2,\dots,k-1}^{Lin}x_{k,d}\right]\right)^2}
\end{cases}$$

$$(3.10)$$

We thus get k different values of optimal μ for which the simplest aggregation method would be to take the mean. However, the value of μ will have a different importance inside each fold. In some folds, the choice of the optimal value[5] of μ leads to a dramatic decrease of the RMSE, whereas in some others, the RMSE is less sensitive to the choice of the optimum. In order to take this into account, for each of the k folds we compute the following metric:

$$\xi_l = \sqrt{E\left[\left(\mathfrak{R}_l - \mathfrak{r}_l^*\right)^2\right]}. \qquad (3.11)$$

Here $\mathfrak{R}_l$ is the vector of the root mean squared errors computed for all the values of μ tested for the fold " l" (about 100), and $\mathfrak{r}_l^*$ is the RMSE value at the optimum.

Thus the metric ξ measures the dispersion of the errors obtained around the optimum. It can be understood as the standard deviation from the minimum RMSE. The larger the value of ξ, the larger the increase of the error when we deviate from the optimum, the more important the choice of this particular value of μ.

We integrate this measure of the importance of the choice of the optimum per fold by computing the final aggregated value of μ as an average of the optimal values obtained from the k-folds, weighted by their associated standard deviation from the minimum:

$$\mu^* = \sum_{l=1}^{k}\mu_l\frac{\xi_l}{\sum_{l=1}^{k}\xi_l}. \qquad (3.12)$$

---

[5] In the sense of the minimization of the pseudo out-of-sample root mean squared error.

Such a weighting is quite intuitive, as our standard deviation to the minimum is strongly related to the notion of standard error.

Note that the procedure described above leads to two OLS fits (one for the linear and one for the non-linear model) per fold, and two final OLS fits with the full data available. Hence, in the case of a 10-fold cross-validation, only 22 matrix inversions are performed, which should be compared with the 100 matrix inversions required in the same case for the Ridge regularization. This difference allows us to anticipate a lower computational time for the LNLM model.

3.3   Evaluation Methodology

We propose to evaluate the interest of the LNLM model using simulations reflecting real-life cases. In our context, the point of modeling is to identify a relation between two variables for which we only have access to noisy observations. Thus we follow the methodology below:

- First simulate a variable X, distributed similarly to stock returns.
- We then associate to this variable a particular reaction function $\phi(X)$ that models the relation between a target and our independent variable.

$$Y = \phi(X). \tag{3.13}$$

- Then, we associate to the output of this function a noise term, thus defining the observed target variable Y, such as:

$$\widetilde{Y} = \phi(X) + \varepsilon. \tag{3.14}$$

- We assume that it is possible to observe the values of X and $\widetilde{Y}$, but that the functional form, as well as the stochastic part of the model, are unknown by an external observer. Putting ourselves in the position of this external observer, we try to fit the mean equation using several modeling techniques.
- Once the fits using these different models are performed, we generate new values of X using the same distribution as before, and see how well the estimated models fit their original target,[6] $\phi(X)$, using these new values of X.

This experiment design thus captures the *out-of-sample predictive power* of each modeling technique, i.e. its ability to tackle the problem of overfitting.

X is drawn from a Student's t distribution with 4 degrees of freedom, as this distribution has been found to appropriately model the distribution of stock returns in a wide range of cases (Platen & Rendek, 2008). For convenience, in the definition of the reaction functions, values drawn outside of the interval $[-6, +6]$ are winsorized.[7]

---

[6]The noise term is not added again in this final step since we want to control for the ability of the models to fit the real underlying response function and not the noise term.

[7]This choice is also made to reflect daily stock returns expected over a -6% to $+6\%$ interval.

Modeling financial markets is often a difficult task because the data used is extremely noisy. Our concern is to test the accuracy of different modeling techniques in the presence of a large amount of noise, thus the values of ε are also to be drawn from a Student's t distribution with 4 degrees of freedom.

We use several base functions to generate different target variables. Our goal is to obtain a representative set of plausible real-life functions. We thus define the thirteen functions below, still in the interval $[-6, +6]$:

$$\phi_1(x) = 0.33x \tag{3.15}$$

$$\phi_2(x) = 0.8 + 0.8x \tag{3.16}$$

$$\phi_3(x) = -2 + 0.75x + 0.2x^2 \tag{3.17}$$

$$\phi_4(x) = 2 + \cos\left(\frac{x}{2}\right) + 0.5x \tag{3.18}$$

$$\phi_5(x) = 0.01\ e^x - 0.1x^2 \tag{3.19}$$

$$\phi_6(x) = 0.1 + 0.1x + 0.02x^2 + 0.03x^3 \tag{3.20}$$

$$\phi_7(x) = 0.1 + 0.1\ \sin(x) - 0.3x \tag{3.21}$$

$$\phi_8(x) = -3 - 0.5x + 0.05x^2 \tag{3.22}$$

$$\phi_9(x) = 0.1 - 0.01x + 0.002x^2 - 0.001x^3 + 0.001x^4 \tag{3.23}$$

$$\phi_{10}(x) = 3 + \tanh(x) + 0.5x \tag{3.24}$$

$$\phi_{11}(x) = -0.4 + 0.5|x|) \tag{3.25}$$

$$\phi_{12}(x) = 0.5\ \sinh(0.01x) - 0.005x^3 \tag{3.26}$$

$$\phi_{13}(x) = 3. \tag{3.27}$$

The different sub-functions and numerical values used to calibrate the base functions have been selected according to several criteria:

- The function must behave smoothly on its interval of definition, reflecting the plausibility of observing such functions in the real-life cases for which the LNLM model is designed. Below are the graphical representations of the functions (Fig. 3.3):

- The slope of the function must be not too sharp. The synthetic observations are created by adding noise to the fitted values, which results in creating a vertical distance between the noisy data and the true data. If we take the extreme case of an infinite slope, a vertical line, any noisy data would be part of the original curve, making the fitting too easy. One can get a good intuitive idea of what is happening here by displaying the noisy observations for the second function (Fig. 3.4):

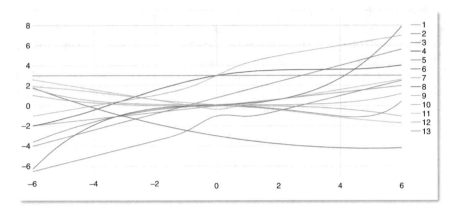

Fig. 3.3 Graphical representations of functions used for simulations

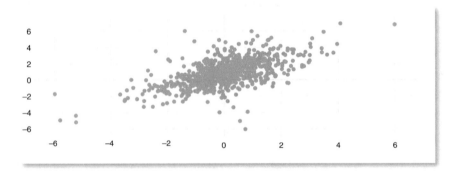

Fig. 3.4 Noisy observations for the second function (slope = 0.8)

Increasing the slope from 0.8 to 3 leads, *ceteris paribus*, to the following noisy data (Fig. 3.5):

Thus, restricting the sharpness of the functions is important to avoid revealing the obvious nature of the underlying functions.

- The sub-functions must be selected in order to represent a wide range of possible functional forms. We particularly take care to include the simplistic constant, linear and quadratic functions, but we also integrate a lot of sub-functions that are not included in the LNLM model itself (these represent roughly half of the functions). This last point is very important if we are concerned with intellectual honesty, since in the real-life cases for which the LNLM model is designed, the true function that links the independent and the target variable is not composed of polynomials.

The resulting noisy target variables are fitted using the following modeling techniques:

- A simple linear model estimated with OLS. The estimation is performed using the python library statsmodels 0.6.1.

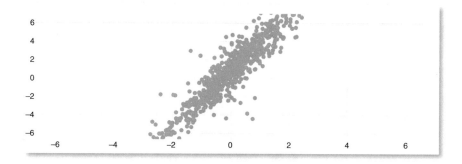

Fig. 3.5 Noisy observations for the second function (slope = 3)

- A polynomial model of the same degree (4) as the LNLM model, estimated with OLS. This model is also estimated with statsmodels.
- A polynomial model of the same degree as the LNLM model, estimated by a Ridge regularization of least squares. The estimation uses the "RidgeCV" function of the library scikit-learn 0.22.2 for python, which allows us to cross-validate the regularization parameter. We search numerically using a 10-fold cross-validation for 13 different values of λ from 1e−8 to 1e+4.
- The LNLM Model, estimated with our variation of stratified k-fold cross-validation. The number of folds used to compute the value of the μ in the LNLM model is set to 10. OLS fits are directly coded in python, using numpy 1.10.4 to invert the covariance matrices of the OLS estimates.
- A naive version of the LNLM Model, estimated with μ set to 0.5. This fit allows us to control for the relevance of the algorithm of choice of μ.
- A non-parametric model, namely the Nadaraya–Watson estimator (again, see Nadaraya (1964), Watson (1964) for a formal definition). The most important parameter of the non-parametric fit, i.e. the bandwidth, is also selected using a cross-validation methodology. The estimation is performed using the kernel regression module of statsmodels.

We also use a different number of observations in the vector X, as with the same simulated noise, it may be much more difficult to identify the underlying function with a small dataset than with a large one. Keeping the same concern for realism, for the vector X we use lengths of 126, 252, 756 and 1,260 observations. These numbers come from the frequent use of rolling windows to estimate polymodels.[8]

In each of the simulations, the data for the values of X used for the estimations, the data for the noise values, and the data of the values of X used for the predictions are generated from different random seeds (from the numpy library).

---

[8] 126 observations corresponds to 6 months of daily data, 252 to 1 year, 756 to 3 years and 1,260 to 5 years.

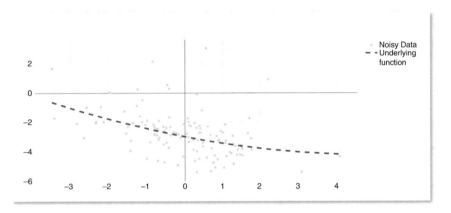

Fig. 3.6 Synthetic noisy data example: Function 8 with 126 observations

To give the reader a graphical impression of the realism and difficulty of the fit, we present below a few examples of the noisy data we generated (Figs. 3.6, 3.7, 3.8, and 3.9):

These plots show the fact that, as in real-life cases, the fits are relatively difficult because of the large amount of noise in the simulated data.

For each of these lengths of X, we run 1,000 simulations with different random seeds for each of the 13 functions defined above, reaching in this way a total of 52,000 simulations.[9]

We compute the root mean squared error of the out-of-sample predictions for all of these cases, which is our indicator of out-of-sample goodness of fit.

We also record the average computation time for each model, composed of the time used to perform the estimations plus the time used to perform the predictions.

3.4 Results

Below we present the tables of results, for each of the four lengths of X.

We first present the summary statistics of the RMSE, that aggregate all the fits for a particular length of X.

We then display the average results for each function, in order to see the ability of each model to fit particular functional forms.

We conclude by presenting the average computation time.

---

[9] 1,000 simulations * 13 functions * 4 vector lengths.

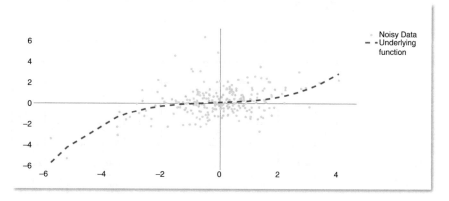

Fig. 3.7 Synthetic noisy data example: Function 6 with 252 observations

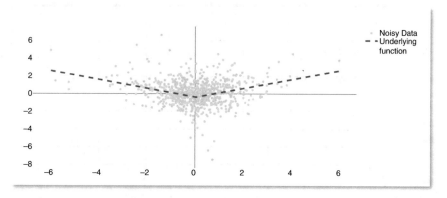

Fig. 3.8 Synthetic noisy data example: Function 11 with 756 observations

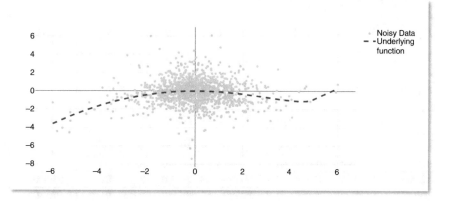

Fig. 3.9 Synthetic noisy data example: Function 5 with 1,260 observations

3.4.1 For 126 Observations (Tables 3.1 and 3.2)

Table 3.1 Summary statistics of the root mean square error, for 126 observations

	Mean	Std	Median	Min	Max
Linear	2.82E−01	1.49E−01	2.70E−01	4.30E−03	1.08E+00
LNLM	3.18E−01	3.21E−01	2.53E−01	8.37E−03	7.66E+00
Naive LNLM	3.51E−01	3.34E−01	2.66E−01	1.76E−02	6.18E+00
Non-Parametric	2.94E−01	1.52E−01	2.77E−01	2.34E−04	2.05E+00
Polynomial by OLS	5.43E−01	6.82E−01	3.22E−01	3.00E−02	1.20E+01
Polynomial by Ridge	5.63E−01	5.34E−01	4.08E−01	2.87E−03	1.01E+01

Table 3.2 Average root mean square error per function fitted, for 126 observations

	Linear	LNLM	Naive LNLM	Non-Parametric	Polynomial by OLS	Polynomial by Ridge
Function #1	1.54E−01	2.15E−01	3.05E−01	2.77E−01	5.13E−01	5.05E−01
Function #2	1.54E−01	2.25E−01	3.05E−01	3.60E−01	5.13E−01	7.93E−01
Function #3	4.11E−01	5.14E−01	5.09E−01	3.84E−01	7.48E−01	9.67E−01
Function #4	3.54E−01	3.58E−01	3.51E−01	3.28E−01	5.13E−01	6.80E−01
Function #5	3.44E−01	3.52E−01	3.51E−01	2.79E−01	5.18E−01	3.96E−01
Function #6	4.60E−01	3.94E−01	3.74E−01	3.96E−01	5.13E−01	4.84E−01
Function #7	1.66E−01	2.30E−01	3.08E−01	2.62E−01	5.15E−01	4.42E−01
Function #8	2.57E−01	2.97E−01	3.22E−01	3.17E−01	5.13E−01	6.48E−01
Function #9	1.95E−01	2.42E−01	3.10E−01	1.93E−01	5.13E−01	2.65E−01
Function #10	3.36E−01	3.82E−01	3.86E−01	3.68E−01	6.14E−01	1.09E+00
Function #11	5.00E−01	4.77E−01	4.33E−01	3.18E−01	5.62E−01	5.24E−01
Function #12	1.74E−01	2.34E−01	3.08E−01	1.84E−01	5.13E−01	2.68E−01
Function #13	1.54E−01	2.17E−01	3.05E−01	1.58E−01	5.13E−01	2.59E−01
Mean	2.82E−01	3.18E−01	3.51E−01	2.94E−01	5.43E−01	5.63E−01
Std	1.26E−01	1.03E−01	6.20E−02	7.80E−02	6.84E−02	2.64E−01
Median	2.57E−01	2.97E−01	3.22E−01	3.17E−01	5.13E−01	5.05E−01
Min	1.54E−01	2.15E−01	3.05E−01	1.58E−01	5.13E−01	2.59E−01
Max	5.00E−01	5.14E−01	5.09E−01	3.96E−01	7.48E−01	1.09E+00

3.4.2 For 252 Observations (Tables 3.3 and 3.4)

Table 3.3 Summary statistics of the root mean square error, for 252 observations

	Mean	Std	Median	Min	Max
Linear	2.49E−01	1.42E−01	2.37E−01	4.43E−03	8.19E−01
LNLM	2.19E−01	1.59E−01	1.90E−01	4.28E−03	3.40E+00
Naive LNLM	2.25E−01	1.39E−01	2.02E−01	1.61E−02	2.75E+00
Non-Parametric	2.30E−01	1.02E−01	2.24E−01	1.61E−04	8.07E−01
Polynomial by OLS	2.87E−01	2.61E−01	2.27E−01	2.33E−02	5.36E+00
Polynomial by Ridge	4.21E−01	3.28E−01	3.29E−01	7.52E−03	5.14E+00

Table 3.4 Average root mean square error per function fitted, for 252 observations

	Linear	LNLM	Naive LNLM	Non-Parametric	Polynomial by OLS	Polynomial by Ridge
Function #1	1.10E-01	1.39E-01	1.69E-01	2.21E-01	2.61E-01	3.75E-01
Function #2	1.10E-01	1.41E-01	1.69E-01	2.75E-01	2.61E-01	5.47E-01
Function #3	3.93E-01	3.92E-01	3.67E-01	2.98E-01	4.35E-01	7.97E-01
Function #4	3.33E-01	2.43E-01	2.33E-01	2.54E-01	2.60E-01	5.10E-01
Function #5	3.22E-01	2.46E-01	2.33E-01	2.24E-01	2.67E-01	2.91E-01
Function #6	4.32E-01	2.54E-01	2.67E-01	2.95E-01	2.61E-01	2.89E-01
Function #7	1.25E-01	1.51E-01	1.74E-01	2.10E-01	2.64E-01	3.27E-01
Function #8	2.25E-01	2.05E-01	1.96E-01	2.50E-01	2.61E-01	4.73E-01
Function #9	1.56E-01	1.69E-01	1.78E-01	1.56E-01	2.61E-01	1.69E-01
Function #10	3.09E-01	2.98E-01	2.71E-01	2.86E-01	3.68E-01	9.62E-01
Function #11	4.77E-01	3.16E-01	3.21E-01	2.52E-01	3.12E-01	3.95E-01
Function #12	1.34E-01	1.55E-01	1.73E-01	1.48E-01	2.61E-01	1.73E-01
Function #13	1.10E-01	1.39E-01	1.69E-01	1.20E-01	2.61E-01	1.66E-01
Mean	2.49E-01	2.19E-01	2.25E-01	2.30E-01	2.87E-01	4.21E-01
Std	1.35E-01	8.08E-02	6.52E-02	5.80E-02	5.45E-02	2.41E-01
Median	2.25E-01	2.05E-01	1.96E-01	2.50E-01	2.61E-01	3.75E-01
Min	1.10E-01	1.39E-01	1.69E-01	1.20E-01	2.60E-01	1.66E-01
Max	4.77E-01	3.92E-01	3.67E-01	2.98E-01	4.35E-01	9.62E-01

3.4.3   For 756 Observations (Tables 3.5 and 3.6)

Table 3.5 Summary statistics of the root mean square error, for 756 observations

	Mean	Std	Median	Min	Max
Linear	2.18E−01	1.41E−01	1.96E−01	2.30E−03	5.47E−01
LNLM	1.29E−01	7.41E−02	1.10E−01	4.22E−03	6.40E−01
Naive LNLM	1.53E−01	8.08E−02	1.39E−01	1.67E−02	4.95E−01
Non-Parametric	1.52E−01	6.37E−02	1.50E−01	7.60E−05	7.02E−01
Polynomial by OLS	1.41E−01	7.18E−02	1.23E−01	1.72E−02	7.95E−01
Polynomial by Ridge	3.08E−01	3.00E−01	1.81E−01	1.43E−02	1.24E+00

Table 3.6 Average root mean square error per function fitted, for 756 observations

	Linear	LNLM	Naïve LNLM	Non-Parametric	Polynomial by OLS	Polynomial by Ridge
Function #1	6.53E−02	7.55E−02	8.19E−02	1.50E−01	1.14E−01	2.71E−01
Function #2	6.53E−02	7.64E−02	8.19E−02	1.83E−01	1.14E−01	3.04E−01
Function #3	3.79E−01	2.98E−01	3.08E−01	2.01E−01	2.95E−01	8.21E−01
Function #4	3.12E−01	1.19E−01	1.74E−01	1.70E−01	1.14E−01	3.53E−01
Function #5	3.00E−01	1.29E−01	1.72E−01	1.48E−01	1.22E−01	1.66E−01
Function #6	4.00E−01	1.19E−01	2.13E−01	1.73E−01	1.14E−01	1.50E−01
Function #7	8.57E−02	8.98E−02	9.07E−02	1.44E−01	1.19E−01	2.28E−01
Function #8	1.97E−01	1.15E−01	1.24E−01	1.68E−01	1.14E−01	3.30E−01
Function #9	1.20E−01	1.01E−01	9.57E−02	1.07E−01	1.14E−01	9.11E−02
Function #10	2.88E−01	2.12E−01	2.15E−01	1.94E−01	2.13E−01	8.83E−01
Function #11	4.59E−01	1.72E−01	2.63E−01	1.72E−01	1.68E−01	2.36E−01
Function #12	9.51E−02	9.11E−02	8.82E−02	1.01E−01	1.14E−01	8.91E−02
Function #13	6.53E−02	7.56E−02	8.19E−02	7.06E−02	1.14E−01	8.71E−02
Mean	2.18E−01	1.29E−01	1.53E−01	1.52E−01	1.41E−01	3.08E−01
Std	1.44E−01	6.46E−02	7.72E−02	3.85E−02	5.50E−02	2.58E−01
Median	1.97E−01	1.15E−01	1.24E−01	1.68E−01	1.14E−01	2.36E−01
Min	6.53E−02	7.55E−02	8.19E−02	7.06E−02	1.14E−01	8.71E−02
Max	4.59E−01	2.98E−01	3.08E−01	2.01E−01	2.95E−01	8.83E−01

3.4.4   For 1,260 Observations (Tables 3.7 and 3.8)

Table 3.7 Summary statistics of the root mean square error, for 1,260 observations

	Mean	Std	Median	Min	Max
Linear	2.09E−01	1.40E−01	1.88E−01	2.94E−03	5.12E−01
LNLM	1.16E−01	6.61E−02	9.58E−02	1.12E−02	4.38E−01
Naive LNLM	1.43E−01	7.97E−02	1.22E−01	2.14E−02	3.83E−01
Non-Parametric	1.37E−01	5.17E−02	1.38E−01	2.95E−03	5.46E−01
Polynomial by OLS	1.26E−01	6.25E−02	1.08E−01	2.77E−02	4.88E−01
Polynomial by Ridge	2.58E−01	2.70E−01	1.40E−01	1.20E−02	1.08E+00

Table 3.8 Average root mean square error per function fitted, for 1,260 observations

	Linear	LNLM	Naïve LNLM	Non-Parametric	Polynomial by OLS	Polynomial by Ridge
Function #1	5.63E−02	7.02E−02	7.14E−02	1.38E−01	9.92E−02	2.28E−01
Function #2	5.63E−02	7.07E−02	7.14E−02	1.64E−01	9.92E−02	2.06E−01
Function #3	3.77E−01	2.84E−01	3.03E−01	1.77E−01	2.83E−01	7.89E−01
Function #4	3.05E−01	1.02E−01	1.68E−01	1.54E−01	9.94E−02	3.22E−01
Function #5	2.92E−01	1.10E−01	1.66E−01	1.34E−01	1.06E−01	1.31E−01
Function #6	3.80E−01	1.01E−01	2.02E−01	1.50E−01	9.92E−02	1.33E−01
Function #7	7.80E−02	8.27E−02	8.02E−02	1.26E−01	1.04E−01	1.90E−01
Function #8	1.89E−01	9.78E−02	1.12E−01	1.45E−01	9.92E−02	2.61E−01
Function #9	1.10E−01	8.87E−02	8.29E−02	9.91E−02	9.92E−02	7.85E−02
Function #10	2.81E−01	1.93E−01	2.07E−01	1.74E−01	1.94E−01	6.78E−01
Function #11	4.52E−01	1.53E−01	2.53E−01	1.53E−01	1.54E−01	1.83E−01
Function #12	8.58E−02	8.19E−02	7.69E−02	9.41E−02	9.92E−02	7.70E−02
Function #13	5.63E−02	6.97E−02	7.14E−02	7.87E−02	9.92E−02	7.52E−02
Mean	2.09E−01	1.16E−01	1.43E−01	1.37E−01	1.26E−01	2.58E−01
Std	1.44E−01	6.17E−02	7.89E−02	3.07E−02	5.52E−02	2.25E−01
Median	1.89E−01	9.78E−02	1.12E−01	1.45E−01	9.92E−02	1.90E−01
Min	5.63E−02	6.97E−02	7.14E−02	7.87E−02	9.92E−02	7.52E−02
Max	4.52E−01	2.84E−01	3.03E−01	1.77E−01	2.83E−01	7.89E−01

3.4.5   Computation Time (Table 3.9)

Table 3.9 Average computation time per model

	126 observations	252 observations	756 observations	1,260 observations	Average
Linear	*1.00E−02*	*1.11E−02*	*8.83E−03*	*1.01E−02*	*1.00E−02*
Naive LNLM	*1.31E−02*	*1.60E−02*	*1.43E−02*	*1.55E−02*	*1.47E−02*
Polynomial by OLS	*2.16E−02*	*2.25E−02*	*1.98E−02*	*1.78E−02*	*2.04E−02*
LNLM	8.88E−01	9.65E−01	8.48E−01	9.00E−01	9.00E−01
Polynomial by Ridge	1.51E+00	1.60E+00	1.62E+00	1.36E+00	1.52E+00
Non-Parametric	1.52E+00	2.60E+00	9.07E+00	1.75E+01	7.67E+00

3.4.6   Interpretations

There are various and interesting conclusions that can be drawn from the previous tables.

First, the most obvious finding is the out-of-sample quality of the fits provided by the LNLM model, which over-perform all the other modeling techniques in terms of Median RMSE in each of the 4 different window lengths, and which is still better in terms of Average RMSE in 3 out of 4 different window lengths.

As expected, we note that, on average, the naive version of the LNLM brings results that are always worse than the LNLM estimated with Stratified K-folds, which gives some credibility to this algorithm and to the relevance of its choice of the non-linearity propensity parameter. Still concerning regularization methods, we also see that it is not clear that Ridge regularization can improve the polynomial model estimated by OLS. However, this can be due to the lack of accuracy of the values of λ evaluated in the cross-validation, since we progress from one λ to another by a factor 10.

Remarkably, the LNLM model does not often beat the other modeling techniques in each particular functional form, it is even beaten most of the time. The other modeling techniques tend to dominate only in particular cases, the most obvious example being the linear model that achieves the best results when the underlying function is indeed linear, i.e. for functions 1, 2 and 13. The emergence of the over-performance of the LNLM model only at the aggregated level is a good sign of its robustness to identify unknown (and various) functional forms.

The kernel estimator achieved good results, especially for small windows. Notably, the standard deviation of its RMSEs is most of the time the smallest one. However, the computation time for this model grows linearly with the number of observations, leading to poor computational time on average.

Apart from this model, all the other computation times are relatively independent of the number of observations. Regularized polynomials using Ridge takes more time than LNLM to be estimated, however, the computation time directly depends on the number of values tested for λ by the cross-validation algorithm. Still, by testing only 13 different values of λ, we have a low level of accuracy for the numerical optimization of this parameter (100 values of μ are tested for LNLM in the current setting). This finding may easily be linked with the number of matrix inversions required for LNLM versus cross-validated Ridge (which is 22 versus 130 in our case), which sheds light on the differences of computational performance. Hence, among the sophisticated methods that are presented, namely the non-parametric fit, the polynomial regularized model, and the LNLM, all using the same principle of cross-validation, the LNLM model appears to be a convincing alternative in terms of computation time.

The performance of the linear model is excellent with a small window of 126 observations, because it is not fooled by the large quantity of noise, unlike the other models, but it becomes worse and worse as the window increases, allowing non-linear techniques to more accurately identify the underlying functional forms.

The inverse phenomenon appears for the polynomial model, which is totally fooled by the noise for small windows, but performs relatively well for large windows, when the underlying function becomes easier to capture, as errors compensate each other more frequently.

For very small or very large windows, outside the scope of the present study, these two models can be interesting, but for the windows that may reasonably be assumed to be used by researchers working on daily financial data, the LNLM model appears to be a convincing alternative.

3.5 Conclusions

We have presented the motivations as well as a complete fitting procedure for the LNLM model, and explained its particular interest compared to other methods of regularization, such as Ridge regression.

The present chapter demonstrates, using a realistic simulation framework, that the LNLM model can successfully reduce over-fitting compared to several alternatives. Of course, there is no guarantee that this will be the case in all possible applications, but it still emphasizes the interest of the model for finance.

Furthermore, the model has the advantage of having a data-driven functional form. Also, the estimated underlying function is always smooth, which can be an advantage for several computational applications, as well as for the realism of the representation of the underlying function (for example, the non-parametric methods can exhibit some disruptions that never occur with LNLM).

The estimations are especially fast, which greatly improves the computation time compared to non-parametric methods, as well as the Ridge-regularized polynomial model (to a lesser extent), which uses recent python libraries, for which the computational time has been optimized. This point is of great importance in the context of big data, which polymodels frequently involve.

All these properties of the LNLM model open a door for applications that the alternative modeling methods may have left closed.

References

Cherny, A., Douady, R., & Molchanov, S. (2010). On measuring nonlinear risk with scarce observations. *Finance and Stochastics, 14*(3), 375–395.

Golub, G. H., Heath, M., & Wahba, G. (1979). Generalized cross-validation as a method for choosing a good ridge parameter. *Technometrics, 21*(2), 215–223.

Hoerl, A., & Kennard, R. (1988). *Ridge regression, in Encyclopedia of Statistical Sciences* (Vol. 8). Wiley.

Moore, A. W. (2001). *Cross-validation for detecting and preventing overfitting.* School of Computer Science Carneigie Mellon University.

Nadaraya, E. A. (1964). On estimating regression. *Theory of Probability and its Applications, 9*(1), 141–142.

Platen, E., & Rendek, R. (2008). Empirical evidence on Student-t log-returns of diversified world stock indices. *Journal of Statistical Theory and Practice, 2*(2), 233–251.

Tibshirani, R. (1996). Regression shrinkage and selection via the lasso. *Journal of the Royal Statistical Society. Series B (Methodological), 58*, 267–288.

Watson, G. S. (1964). Smooth regression analysis. *Sankhya: The Indian Journal of Statistics, Series A, 26*(4), 359–372.

Zhang, J. (2019). *Statistical arbitrage based on stock clustering using nonlinear factor model* (Doctoral dissertation). State University of New York at Stony Brook.

Zou, H., & Hastie, T. (2005). Regularization and variable selection via the elastic net. *Journal of the Royal Statistical Society: Series B (Statistical Methodology), 67*(2), 301–320.

Chapter 4
Predictions of Market Returns

Abstract We propose a Systematic Risk Indicator derived from a polymodel estimation. Polymodels allow us to measure the strength of the links that a stock market maintains with its economic environment. We show that these links tend to be more *extreme* before a market crisis, confirming the well-known increase of correlations while proposing a more subtle perspective. A fully automated and successful trading strategy is implemented to assess the interest of the signal, which is shown to be strongly significant, both from an economic and statistical point of view. Results are robust across different time-periods, for various sets of explanatory variables, and among 12 different stock markets.

Keywords Polymodel theory · Systematic risk · Financial crisis · Speculative bubble · Stock market crash · Trading strategy · LNLM modeling · Market timing · Drawdown prediction

4.1 Introduction

In this chapter, we predict the market returns of 12 main stock indexes.

Our approach takes root in the wide literature dedicated to market crisis anticipation. Quite often, this literature relies on identifying some herding behavior.

Sornette (see Johansen et al. (2000) and Sornette (2009)) has been a pioneer in the anticipation of stock market crashes with his LPPL model (also well described in Geraskin and Fantazzini (2013)). Essentially, this model describes the evolution of the stock market price during a speculative bubble. The market dynamic is modeled using a combination of a power law, and some decreasing periodic oscillations. The power law in the price is the result of the actions of a group of rational, well-informed traders, who invest in a speculative bubble because the risk of stock market crash is compensated by the returns produced by the acceleration of the price increase. The oscillations are produced by the random actions of a group of noise traders, who exhibit some herding behavior, for example by being driven away from the rational price by an overreaction to the news. The decrease in the oscillation is because when the power law strengthens, approaching its final singularity, all the market participants identify the bubble and behave as rational traders. Note that such a behavior by market participants implies an increasing level of mimicry among traders' actions while approaching the stock market crash.

Anticipation of a financial crisis also often relies on identifying increasing correlations among assets composing the markets. Patro et al. (2013) propose a very simple systemic risk indicator by measuring increasing correlations among stock returns of banks. This kind of approach has numerous variations, see Silva et al. (2017) for a survey. Some other authors do not restrict their approach to identifying correlations among banks. For example, Douady and Kornprobst (2018) developed an approach based on the structure of the correlation matrix of the stocks defining a market. Zheng et al. (2012) use an indicator based on the link among sectorial sub-components of the market index to predict the market itself. Harmon et al. (2011) show that the internal herding of the market can also be captured by co-movement.

But these increasing correlations also appear among the markets themselves, which tend to behave as one during a financial crisis (Junior & Franca, 2012). This point is especially important, because it highlights the fact that external causes also matter, along with internal dynamics. Ye and Douady (2019) use 200 external factors to estimate a non-linear polymodel, with successful results in deriving a systemic risk indicator. Their indicator is based on the sum among elementary models of a goodness of fit indicator, the R^2. Our approach also relies on assessing the links between a stock market index and its environment using a non-linear polymodel. However, although our systemic risk indicator is focused on the goodness of fit, we do not make any assumptions about how the goodness of fit should evolve during a financial crisis. In particular, we do not assume that the links between the market and its environment strengthen during the crisis, and we will see that this strengthening is only a part of the phenomenon's characteristics.

In this chapter, we first present the structure of the data that we use to estimate the non-linear polymodel. Next, we propose an economically agnostic and point-in-time systemic risk indicator. We also analyze the indicator with respect to various aspects. In a third part, we simulate a trading strategy, in order to confront the indicator with its most natural real-world implementation: trying to beat the market. The fourth part is dedicated to performing a wide set of robustness tests. Finally, we summarize the findings and limits of the chapter in a conclusion section, which is followed by the references.

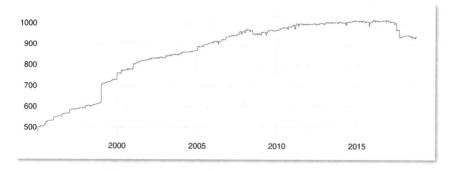

Fig. 4.1 Number of variables in the factor set over time

4.2   Data

A large majority (96%) of the data used is provided by Bloomberg, in the form of daily returns.[1,2] Since all the data is market data, it is point-in-time. The data includes the returns of the 12 stock indexes that are the target variables of our polymodels, and a factor set that includes 1,134 explanatory variables. The dataset is defined between 1995-01-02 and 2018-10-16, although some variables are defined only for a part of this period. Thereafter the number of variables available over time in the factor set is shown (Fig. 4.1):

The target stock indexes are the following:

- S&P 500 Index
- Euro STOXX 600 Index
- FTSE 100 Index
- Tokyo Stock Exchange Price Index (TOPIX)
- DAX Index
- CAC 40 Index
- Swiss Market Index
- IBEX 35 Index
- OMX Copenhagen Index
- OMX Helsinki Index
- OMX Stockholm 30 Index
- MSCI Italy Index

The factor set is designed with the available data in order to reach a near-complete representation of the world financial markets. However, we allow an important bias in favor of the equity indices, as the target variables are themselves equities. This

---

[1] Daily returns are computed using total return indexes when they are available, otherwise we simply use the close price.

[2] The remaining data comes from various sources, essentially being FTSE Equity indices and J.P. Morgan Corporate Bonds indices.

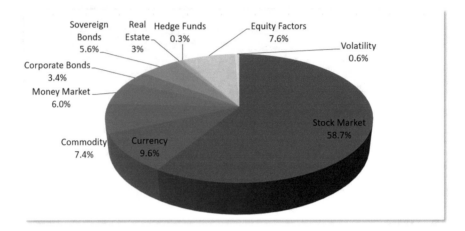

Fig. 4.2 Repartition in terms of asset classes of the factor set

Table 4.1 Repartition of region-specific factors	World	60.41%
	World ex-USA	4.88%
	Developed markets	6.17%
	Emerging markets	4.88%
	Europe	5.91%
	Asia/Pacific	5.40%
	Africa & Middle East	4.37%
	South America	4.11%
	North America	3.86%
	Total	100.00%

choice is also motivated by the broader number of themes covered by this asset class (regional as well as country indices, sectorial indices, factor indices...). Below is the repartition in terms of asset classes of the factor set (Fig. 4.2):

Our concern for the representativeness of the factor set also incorporates the question of geographical coverage. In order to address this question from different perspectives, we select some region-specific factors as well as some country-specific factors.

Region-specific factors include several geographical divisions. At the most inclusive level, we have some World or World ex-USA factors. The simplest division is then between Developed and Emerging Markets. Finally, the region-specific factors also include some continental factors. The region-specific factors represent 34% of the factor set, all the other factors being country-specific. The repartitions of the region-specific factors are shown in the Table 4.1 below:

Country-specific factors thus represent 66% of the factor set. Below is their repartition (Table 4.2):

Table 4.2 Repartition of country-specific factors

United States of America	10.34%
China	5.50%
Japan	3.76%
Italy	2.82%
Swiss	3.09%
Australia	3.36%
Malaysia	2.68%
United Kingdom	3.09%
Egypt	2.55%
Indonesia	2.42%
New Zealand	2.42%
Russia	2.42%
Taiwan	2.68%
Austria	2.28%
France	2.55%
Korea	2.28%
Poland	2.28%
Spain	2.28%
Belgium	2.15%
Germany	2.42%
Hong Kong	2.15%
Mexico	2.15%
Netherlands	2.15%
Brazil	2.01%
Saudi Arabia	2.15%
South Africa	2.01%
Turkey	2.01%
United Arab Emirates	2.01%
Canada	2.68%
Norway	1.88%
Sweden	1.88%
Colombia	1.74%
Denmark	1.74%
Greece	1.74%
Thailand	1.74%
India	1.61%
Argentina	1.48%
Nigeria	1.21%
Ukraine	1.21%
Jamaica	0.67%
Banglasdesh	0.27%
Iran	0.13%
Total	100.00%

4.3 Systemic Risk Indicator

4.3.1 Model Estimation

For each of the stock markets considered, the methodology to derive our systemic risk indicator is the same.

We first estimate a polymodel:

$$\{Y = \varphi_i(X_i) \quad \forall i\}. \tag{4.1}$$

The polymodel is estimated using the LNLM model, as described in Chap. 3:

$$LNLM(X) \overset{\lrcorner}{=} \bar{y} + \mu \sum_{h=1}^{u} \widehat{\beta}_h^{NonLin} H_h(X) + (1 - \mu)\widehat{\beta}^{Lin} X + \varepsilon$$

$$\Rightarrow \{Y = LNLM_i(X_i) \quad \forall i\}. \tag{4.2}$$

The polymodel is estimated on a rolling basis, using a 5-year window. Although the data is available at a daily frequency, we only re-perform the estimation each month, due to our computational resources. We transform the daily returns in order to get some monthly returns,[3] which may exhibit clearer patterns. The factor set of explanatory variables is lagged by one month, in order to give us a clear direction for causality (as a positive side effect, it also makes sure that we avoid the use of future information due to the asynchronicity of international markets):

$$\{Y_t = LNLM_{i,t}(X_{i,t-1}) \quad \forall i\}. \tag{4.3}$$

Once the estimation is done for a given date, we compute a goodness of fit indicator, the root mean square error, for each of the elementary models that compose the polymodel. This indicator captures the strength of the link between a given factor and the predicted stock market. In order to remove the most insignificant factors, we filter the polymodel to keep only the lowest 80% RMSEs.

4.3.2 Systemic Risk Indicator Definition

The distribution of the RMSEs is a simple representation of the links that the market maintains with its whole economic environment. It is similar to having a distribution of the correlations, however, it is a more general representation since it allows for the links to be non-linear, thanks to the use of polymodels. The foundation of our

---

[3]This is done by performing a cumulative product of (1 + return) inside a 21-day rolling window. The resulting data is monthly returns available at a daily frequency.

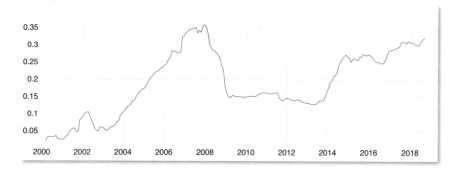

Fig. 4.3 Average Hellinger distance to the Pre-Crisis Distribution, over time

indicator is the hypothesis that a specific distribution of the RMSEs exists before the crisis.

The distributions of the RMSEs are obtained using a simple kernel density estimation (see Parzen, 1962; Rosenblatt, 1956). The bandwidth selection is done according to Silverman's rule of thumb (Silverman, 1986).

First, we only select the distributions of RMSEs that are followed by a market drawdown in the following 3 months. Our representation of the pre-crisis distribution is a rolling average of these distributions, weighted by the square of the 3 months' returns that follows (we assume that the larger the following drawdown, the more representative the distribution). The rolling average follows an exponential weighting scheme, with a half-life of 10 years, that allows us to slightly overweight the most recent distributions, while very old representations of the pre-crisis distribution are taken into account with less and less importance. Of course, this rolling representation is shifted properly so that at each point in time, it does not use any future information.

The 'normal times' distribution is represented by the average of all RMSE distributions, whatever the sign of the following 3 months' return of the market.

We thus obtain a representation of the pre-crisis distribution and a representation of the normal times distribution for each of the 12 stock indexes we analyze.

Once this representation of the pre-crisis distribution is obtained, we compute the Hellinger distance between this distribution and the current distribution of the RMSE. The smaller the Hellinger distance, the closer the market is to a pre-crisis situation. In Fig. 4.3 below, we display the evolution of the average Hellinger distance to the Pre-Crisis Distribution:

The pattern drawn by the Hellinger distance clearly reflects the speculative bubble that preceded the financial crisis of 2008, as well as its burst. Assuming that this pattern is indeed predictive, then a decrease in the distance would be a negative signal about future market returns, and reciprocally.

Our Systematic Risk Indicator simply reproduces this reasoning, being defined by the sign of the variation of the Hellinger distance:

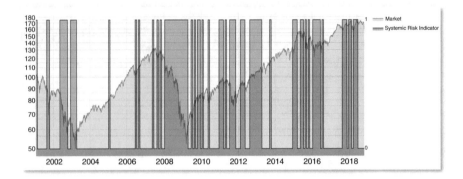

Fig. 4.4 Systematic Risk Indicator and Stoxx Europe 600

$$SRI_t \triangleq \begin{cases} 1 & if \ \mathcal{H}\left(\mathcal{R}_t^{Current}, \mathcal{R}_t^{Pre-Crisis}\right) < \mathcal{H}\left(\mathcal{R}_{t-1}^{Current}, \mathcal{R}_{t-1}^{Pre-Crisis}\right) \\ 0 & if \ \mathcal{H}\left(\mathcal{R}_t^{Current}, \mathcal{R}_t^{Pre-Crisis}\right) \geq \mathcal{H}\left(\mathcal{R}_{t-1}^{Current}, \mathcal{R}_{t-1}^{Pre-Crisis}\right) \end{cases} \qquad (4.4)$$

Here, " $\mathcal{R}_t^{Current}$ " is the distribution of the RMSE at the current date "t", " $\mathcal{R}_t^{Pre-Crisis}$ " is the last representation of the pre-crisis distribution of the RMSE available at date "t" and " $\mathcal{H}(\ )$ " is the Hellinger distance.

The Hellinger distance is defined in our discrete framework as:

$$\mathcal{H}\left(\mathcal{R}_t^{Current}, \mathcal{R}_t^{Pre-Crisis}\right) \triangleq \frac{1}{\sqrt{2}} \sqrt{\sum_{i=1}^{n} \left(\sqrt{\mathcal{R}_{i,t}^{Current}} - \sqrt{\mathcal{R}_{i,t}^{Pre-Crisis}}\right)^2} \qquad (4.5)$$

Here, " $\mathcal{R}_{i,t}^{Current}$ " is the value of the RMSE at the current date "t" for a given factor "i" and " $\mathcal{R}_{i,t}^{Pre-Crisis}$ " is the value of the RMSE for the factor "i" drawn from the last representation of the pre-crisis distribution of the RMSE available at date "t".

As an example, we display[4] below the indicator in the case of the Stoxx Europe 600 (Fig. 4.4):

Although the indicator is not perfect, at first glance it seems to be activated during most of the drawdowns of the market, even if there are some false positives.

4.3.3   Primary Analysis

If our main test for the predictive power of our Systemic Risk Indicator is to implement a trading strategy, we first conduct a simple analysis to better understand the properties of the indicator.

---

[4]For convenience, all the markets compounded returns are displayed with a base 100 at the beginning of the sample (01/12/2001), in log-scale.

Table 4.3 Summary of a logistic regression of the direction of the future market returns on the systemic risk indicator

	a. Full Sample	b. Subprime Crisis Period (2007–2009)	c. After Subprime Crisis Period (2010–2018)
DAX Index	−4.08171 (***)	−3.870322 (***)	−2.629115 (***)
HEX Index	−7.097665 (***)	−9.184638 (***)	−6.63464 (***)
IBEX Index	−5.653079 (***)	−7.82105 (***)	−3.826911 (***)
KAX Index	−9.003569 (***)	−6.395333 (***)	−5.249408 (***)
MXIT Index	−10.995961	−11.167755 (***)	−1.806702 (*)
NCAC Index	−5.725893 (***)	−7.115673 (***)	−0.018437 ()
OMX Index	−1.126313 ()	−1.793216 (*)	0.137003 ()
SMI Index	−7.651904 (***)	−8.600267 (***)	−2.39497 (**)
SPX Index	−9.898438 (***)	−7.752158 (***)	−10.863697 (***)
SXXP Index	−9.53557 (***)	−9.376226 (***)	−3.017459 (***)
TPX Index	−17.485469 (***)	−5.240356 (***)	−11.043662 (***)
UKX Index	−5.492477 (***)	−10.047277 (***)	1.483732 ()
Mean	**−7.812337 (***)**	**−7.363689 (***)**	**−3.822022 (***)**
Median	−7.374785	−7.786604	−2.823287
Std	4.112063	2.691223	4.024313
Min	−17.485469	−11.167755	−11.043662
Max	−1.126313	−1.793216	1.483732

To begin with, we display below the t-statistics obtained by a logistic regression estimated with maximum likelihood (Table 4.3). Since the Pre-Crisis distribution that we identified in the construction of the indicator is defined as being followed by a negative 3-month market return, our variable of interest is a dummy representing the sign of the 3-month future return. The explanatory variable is, of course, the Systemic Risk Indicator itself:

Full sample, the Systemic Risk Indicator is highly significant in 11 markets out of 12, which is a good sign of its predictive power.

Since the indicator is designed precisely to anticipate the crisis, we also conduct some sub-sample regressions, in order to compare the performance of the indicator during the 2008 subprime crisis, and in the decade that followed, which is calmer. The results are significant at the 1% level of p-value for all the markets except OMX, which exhibit only a 7% p-value in the crisis period. Although the indicator is overall less significant in the second period, it continues to attest to a high level of predictive power.

Table 4.4 below displays the information coefficients, which should resonate with the references of the practitioners. We compute both the IC[5] with the sign of the

---

[5] The "Information Coefficient", defined as the correlation between the signal and the future returns.

Table 4.4 Information coefficients of the systemic risk indicator

	DAX Index (%)	HEX Index (%)	IBEX Index (%)	KAX Index (%)	MXIT Index (%)	NCAC Index (%)	OMX Index (%)	SMI Index (%)	SPX Index (%)	SXXP Index (%)	TPX Index (%)	UKX Index (%)
Future returns sign	−6.06	−10.56	−8.40	−13.43	−16.40	−8.51	−1.67	−11.39	−14.79	−14.22	−26.29	−8.16
Future returns value	−4.80	−0.95	−7.40	−4.81	−14.08	−9.47	1.63	−8.40	−9.34	−13.42	−27.82	−6.71

	DAX Index	HEX Index	IBEX Index	KAX Index	MXIT Index	NCAC Index	OMX Index	SMI Index	SPX Index	SXXP Index	TPX Index	UKX Index
DAX Index		57%	63%	55%	56%	79%	72%	61%	70%	76%	37%	57%
HEX Index	57%		40%	36%	27%	42%	43%	37%	41%	49%	16%	32%
IBEX Index	63%	40%		48%	60%	67%	64%	67%	64%	68%	39%	65%
KAX Index	55%	36%	48%		50%	65%	62%	61%	58%	64%	57%	61%
MXIT Index	56%	27%	60%	50%		72%	53%	55%	61%	63%	48%	66%
NCAC Index	79%	42%	67%	65%	72%		73%	65%	72%	81%	49%	73%
OMX Index	72%	43%	64%	62%	53%	73%		66%	65%	75%	44%	60%
SMI Index	61%	37%	67%	61%	55%	65%	66%		72%	76%	49%	67%
SPX Index	70%	41%	64%	58%	61%	72%	65%	72%		76%	49%	69%
SXXP Index	76%	49%	68%	64%	63%	81%	75%	76%	76%		53%	69%
TPX Index	37%	16%	39%	57%	48%	49%	44%	49%	49%	53%		53%
UKX Index	57%	32%	65%	61%	66%	73%	60%	67%	69%	69%	53%	

Fig. 4.5 Correlation matrix of the systemic risk indicator

future 3 months' returns, and with their values. Interestingly, predicting the sign of the future returns is not enough in all the cases, and the value of the IC decrease is important for the HEX and the KAX Index, for example.

These differences of IC between sign and value can be explained by having some false positives/negatives that precede very large market returns.

To conclude this primary analysis, we consider the correlation matrix among indicators. The matrix shows high levels of correlations among indicators, with an average correlation of 58% (diagonal excluded). Such a configuration asks if the Systemic Risk Indicator adapts to each of the markets, or if it just predicts a kind of "average market". Recalling that the indicator predicts the sign of the next 3 months' market returns, we compare the correlation matrix of these signs to the correlation matrix of the indicator (Figs. 4.5 and 4.6).

Comparing these two matrices leads to the conclusion that the correlation among indicators just reflects the correlations among market returns. The average correlation of the second matrix is similar to that of the first matrix: 57%.

We should also consider that some indicators which are particularity lowly correlated, such as those of HEX and TPX, well perform individually in terms of t-stats. Hence, we can reasonably consider that our Systemic Risk Indicator is adaptive to the specificities of each of the different markets.

	DAX Index	HEX Index	IBEX Index	KAX Index	MXIT Index	NCAC Index	OMX Index	SMI Index	SPX Index	SXXP Index	TPX Index	UKX Index
DAX Index		61%	64%	53%	60%	75%	63%	64%	56%	80%	44%	67%
HEX Index	61%		51%	52%	48%	59%	61%	56%	51%	66%	29%	56%
IBEX Index	64%	51%		42%	63%	69%	52%	50%	48%	67%	35%	60%
KAX Index	53%	52%	42%		47%	55%	56%	54%	52%	61%	40%	54%
MXIT Index	60%	48%	63%	47%		69%	48%	54%	53%	64%	34%	58%
NCAC Index	75%	59%	69%	55%	69%		66%	66%	60%	84%	43%	72%
OMX Index	63%	61%	52%	56%	48%	66%		60%	56%	70%	37%	60%
SMI Index	64%	56%	50%	54%	54%	66%	60%		57%	75%	43%	65%
SPX Index	56%	51%	48%	52%	53%	60%	56%	57%		63%	42%	59%
SXXP Index	80%	66%	67%	61%	64%	84%	70%	75%	63%		47%	75%
TPX Index	44%	29%	35%	40%	34%	43%	37%	43%	42%	47%		43%
UKX Index	67%	56%	60%	54%	58%	72%	60%	65%	59%	75%	43%	

Fig. 4.6 Correlation matrix of the signs of the 3 months' market returns

4.3.4 Roots of the Predictive Power

In order to better understand the patterns that are used by the Systemic Risk Indicator to predict the future drawdowns, we conduct below an aggregated analysis of the Pre-Crisis and of the Normal Times distributions of RMSEs.

Instead of using rolling averages for these two distributions, we directly compute full sample averages of the distributions. These full sample average distributions are then averaged again among markets, so that we obtain a single, representative distribution for both the normal and the pre-crisis case.

These representative distributions[6] allow us to understand what happens before the crisis occurs:

Figure 4.7 seems to show a simultaneous increase of some of the RMSEs and a decrease of some others. If we compute the difference of densities of these two distributions, the phenomenon becomes even more obvious (Fig. 4.8):

Above, we separate the density differences into three areas, that correspond roughly to the left tail, the center of the distribution, and the right tail. In the center of the distribution it is clear that there is a decrease of density, which means that the quantity of factors that are usually moderately correlated to the stock index decreases before the crisis. We note that there is an increase in the left tail, which corresponds

---

[6]Distributions are always normalized so that their integral is equal to 1, in order to remove the undesirable effect of a changing number of variables in the factor set over time.

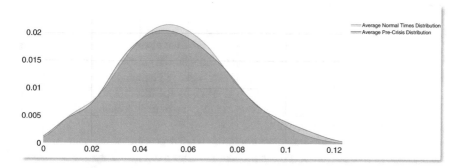

Fig. 4.7 Representative RMSE distributions

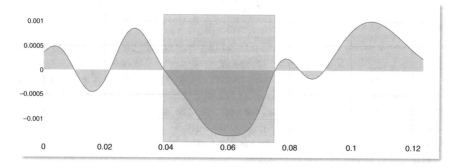

Fig. 4.8 Differences of previous densities between Normal Times and Pre-Crisis distributions

to the expected increase of correlations well documented in the literature. But surprisingly, most of the density that disappears in the center appears in the right tail, which implies that the de-correlation of some factors which are usually correlated is an important characteristic of the pre-crisis time. More precisely, 69.7% of the density that disappears in the center goes to the right tail and only 30.3% to the left tail. These findings, based on the average representative distributions among markets, also seem to be true for each of the markets, in which most of the time the right and the left fat tail coexist, although the right fat tail seems to occur more frequently (see Appendix: Representative RMSE Distributions per Stock Index).

4.4 Trading Strategy

Our principal assessment of the performance of the Systemic Risk Indicator is through a simulated trading strategy.

4.4.1   Methodology

The simulated trading strategy is extremely simple: we always get the returns of the market, but with a variable leverage, equal to 0.5 when the risk is off, and 2 when the risk is on. Since there is a wide range of leveraged ETFs available nowadays, we assume no transaction costs.[7] This is also justified by the fact that since the Systemic Risk Indicator is only computed once a month, the potential change of leveraged ETF also only occurs once a month.

For each of the stock indexes, we compute the Sharpe ratio of the market, and of the market timed portfolio. We then compute the average Sharpe ratio among markets, and the average Sharpe ratio of market timed portfolios. The percentage of improvement of this average Sharpe is the central number of our aggregated analysis.

We assess the significance of this number by computing its p-value. The p-value is generated as follows:

- For each market, we compute the sign of the variation of the Hellinger distances, as we do for the Systemic Risk Indicator.
- We shuffle these signs over the time period of the sample, so that they are present in the same proportions as in the SRI but at different points in time.
- Based on the shuffled indicator, we compute the performance of a new market timed portfolio, using the same trading strategy.
- We then compute the metric we are interested in, i.e. the improvement of the average Sharpe ratio among markets.
- We repeat this procedure 1,000 times, so that we get a distribution of random average Sharpe improvements.
- The distribution is smoothed, using a kernel density estimation, in order to avoid some threshold effects in the p-value computation.
- Given this distribution, we simply compute the probability of randomly obtaining the level of improvement we get originally.

Note that this procedure eventually only evaluates the *relevance of the moment* when the signal appears and makes sure that the performance is unlikely to come from random chance, or from a particular design of the trading strategy.

4.4.2   Results

Table 4.5 summarizes the results of the trading strategy, by presenting the Sharpe ratios of each market and their improvements:

---

[7] Of course, the portfolio returns used thereafter to compute Sharpe ratios are always excess returns, and thus can be considered as an implicit assumption of funding costs to finance the positions of the portfolio.

Table 4.5 Improvements of the Sharpe ratios obtained from the trading strategy

	Market	Market Timed Portfolio	Improvement (%)
DAX Index	0.12	0.31	148
HEX Index	0.20	0.25	21
IBEX Index	0.19	0.45	141
KAX Index	0.38	0.73	89
MXIT Index	0.02	0.30	1,275
NCAC Index	0.09	0.47	394
OMX Index	0.28	0.53	88
SMI Index	0.14	0.50	267
SPX Index	0.35	0.68	95
SXXP Index	0.17	0.53	205
TPX Index	0.00	0.35	12,381
UKX Index	0.21	0.50	140
Average	**0.18**	**0.47**	**157**

The performance of each of the markets is improved by the trading strategy, and the increases of the Sharpe ratios are economically significant.

The p-value of the improvement of the average[8] computed with the methodology mentioned previously is 0.06%, confirming that the performance is also statistically significant.

We display below the distribution of the random performances (Fig. 4.9):

Out of 1,000 simulated random portfolios, only one is better than the proposed portfolio. This distribution also emphasizes the fact that beating the market is extremely difficult, since its average is -14.62%.

Note that on average, the market timed portfolios are hedged 41% of the time, which is a good sign of non-overfitting. Indeed, since the leverage is re-decided each month, there are a lot of different bets that are done over time. Compared to an indicator that only bets against the market a couple of times over 20 years of simulations, it considerably decreases the chances of cherry-picking a crisis indicator.

Staying with the previous example, we display below the performance of the strategy for the Stoxx Europe 600[9] (full results are available in Appendix: Market Timed Portfolio and Systemic Risk Indicator per Stock Index) (Fig. 4.10):

---

[8] Computing the average improvement instead of the improvement of the average would have been more natural. However, it is obvious that since the base Sharpe ratios are very small for some markets, their improvements are extremely high. The average of the improvements is thus contaminated by these outliers, which leads this number to become meaningless.

[9] In this graph, the Market and the Market Timed Portfolio returns are standardized so that they have the same volatility.

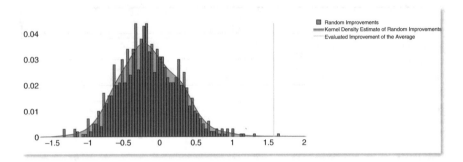

Fig. 4.9 Distribution of randomly generated improvements of the average Sharpe

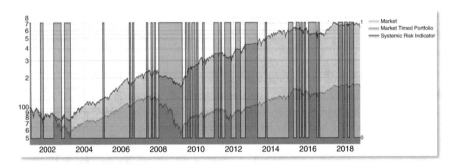

Fig. 4.10 Stoxx Europe 600: Market Timed Portfolio vs Market Portfolio (equal volatility)

4.5   Robustness Tests

In the previous developments, we already saw that frequent changes of the indicator are a good protection against overfitting, and that the indicator is still predictive in sub-samples. We complement these findings with a series of robustness tests that assesses the sensitivity of the performance of the trading strategy to the design we made.

For each of the robustness tests, we present the variation of the significance of the trading strategy in statistical terms (p-value) and economical terms (improvement of the average Sharpe ratio, number of indices for which the strategy beat the market).

4.5.1   Sensitivity to the Noise Filter

Recall that we removed the noisy factors from the RMSE distribution using a very simple filter that only keeps the top X% of the best explanatory variables in terms of RMSE. This parameter was originally set to 80%, in order to keep most of the factors.

Table 4.6 Sensitivity of the improvement of the average Sharpe ratio to the noise filter

Filter quantile (%)	Improvement of the average (%)	p-value (%)	Percentage of positive improvements (%)
100	61.13	7.52 (*)	100.00
95	138.78	0.01 (***)	100.00
90	152.77	0.00 (***)	100.00
85	148.82	0.01 (***)	100.00
80	157.49	0.06 (***)	100.00
75	153.45	0.04 (***)	100.00
50	92.54	0.61 (***)	100.00
25	58.08	3.16 (**)	100.00
10	−0.86	33.49 ()	58.33

We can see in the Table 4.6 that this choice seems well justified:

The first finding that can be extracted from Table 4.6 is that the performance of the trading strategy is robust to the choice of a particular level of filtering, except in the extreme regions.

Secondly, our approach has been to select a reasonable value for the filter. Indeed, it is possible to reach a higher level of historical performance using the 99% quantile, however, such a choice would have seemed dangerous with regard to the performance of the 100% quantile. Some of the variables in between seem to introduce a lot of noise in the RMSE distribution, and the frontier between a relevant and a noisy variable is quite thin in this region.

As a third point, we would like to mention that the vast majority of the factor set seems to be important for the crisis prediction. This emphasizes the need to select a large factor set to describe the entire economic environment of the stock indices we are predicting.

4.5.2 Sensitivity to Future Returns Windows

The indicator is designed so that it predicts the sign of the 3-month future returns of the markets. This choice has been made following some practitioners' heuristics, 1 month being usually considered as too noisy for the crisis prediction, and the months following the next 3 months being considered as less predictable. Our sensitivity analysis on this parameter confirms this opinion, while showing the clear robustness of the trading strategy to this parameter (Table 4.7).

4.5.3 Sensitivity to Half-life

The Pre-Crisis RMSE distribution is defined on a rolling basis, using an exponential weighting scheme of the past observations. The speed at which the current

Table 4.7 Sensitivity of the average improvement of the Sharpe ratio to the future returns window

Future returns window	Improvement of the average (%)	p-value (%)	Percentage of positive improvements (%)
1 Month	117.11	0.12 (***)	83.33
2 Months	131.78	0.10 (***)	100.00
3 Months	157.49	0.06 (***)	100.00
6 Months	147.46	0.00 (***)	100.00
9 Months	139.71	0.00 (***)	100.00
12 Months	115.93	0.00 (***)	100.00

Table 4.8 Sensitivity of the average improvement of the Sharpe ratio to the half-life of the exponential window

Half-life of the exponential window	Improvement of the average (%)	p-value (%)	Percentage of positive improvements (%)
20 Years	149.82	0.11 (***)	100.00
15 Years	152.46	0.10 (***)	100.00
10 Years	157.49	0.06 (***)	100.00
5 Years	161.12	0.00 (***)	100.00
3 Years	154.84	0.00 (***)	100.00
1 Year	96.74	0.41 (***)	100.00

representation of the Pre-Crisis distribution 'forgets' the past events is controlled by the half-life of this exponential window. Our choice has been to set it to 10 years, in order to have a representation that takes into account very old events while focusing a bit more on the present.

The table below shows that this is a good choice (even if it is not the best). Also, the robustness of the strategy performance to this parameter is still quite good (Table 4.8).

4.5.4 Sensitivity to Rolling Window Length

The polymodel that we use to compute the RMSEs is itself estimated on a rolling basis.

We can see below that the length of the rolling window used for this estimation is a parameter that matters (Table 4.9):

Although the improvement of the average and the p-value stays significant in all cases, it is to a less extent than in other robustness tests, both economically and statistically. The quality of the estimations is thus particularly tied to this parameter: a shorter window includes less data points, and make the fit, along with its goodness of fit indicator, less reliable, while a longer window estimates a model that evolves too slowly, and is thus unable to efficiently capture the relatively fast dynamics of the

Table 4.9 Sensitivity of the average improvement of the Sharpe ratio to the polymodels estimation window

Polymodel estimation window	Improvement of the average	p-value	Percentage of positive improvements
6 Years	27.09	9.01 (*)	91.67
5 Years	157.49	0.06 (***)	100.00
4 Years	56.84	8.32 (*)	75.00
3 Years	78.38	1.24 (**)	91.67

market. Considering this reasoning and the results above, the 5-year window then seems quite justified.

However, especially since we are measuring the performance of a trading strategy, it is important to make sure that the construction of a robust trading strategy is possible based on our Systemic Risk Indicator. Hence, in order to limit the risk of arbitrarily picking a window parameter, we recompute the indicator and the performance of the strategy using an average of the RMSEs estimated using a 6-year, a 5-year, a 4-year, and a 3-year rolling window. Computing the indicator using this average RMSE is a simple way to make it more robust to the choice of a particular window.

The result is that this "window robust strategy" still performs quite well. The improvement of the average is 95%, a figure associated with a p-value of 0.16%. In this version of the strategy, previous figures stay comparable, with an average random improvement of -12.63%, an average hedging 35.96% of the time within the sample, and 11 stock indexes out of 12 that over-perform the market. Ultimately, even diluted by weaker versions of itself, our trading strategy delivers strong results (Table 4.10):

Table 4.10 Improvements of the Sharpe ratios for a "window robust strategy"

	Market	Market timed portfolio	Improvement (%)
DAX Index	0.12	0.06	−49
HEX Index	0.20	0.29	41
IBEX Index	0.19	0.33	73
KAX Index	0.38	0.76	99
MXIT Index	0.02	0.13	502
NCAC Index	0.09	0.32	235
OMX Index	0.28	0.30	7
SMI Index	0.14	0.34	153
SPX Index	0.35	0.62	76
SXXP Index	0.17	0.43	149
TPX Index	0.00	0.24	8,311
UKX Index	0.21	0.40	93
Average	**0.18**	**0.35**	**95**

Table 4.11 Sensitivity of the average improvement of the Sharpe ratio to the asset class

Asset class	Improvement of the average (%)	p-value (%)	Percentage of positive improvemsents (%)
Stock markets	104.89	0.33 (***)	91.67
Equity factors	29.99	12.55 ()	66.67
Currencies	67.51	1.47 (**)	83.33
Commodities	71.20	0.62 (***)	100.00
Corporate bonds	33.79	9.87 (*)	83.33
Sovereign bonds	66.99	0.72 (***)	91.67
Money market	−32.26	62.51 ()	16.67

4.5.5 Asset-Class Specific Performances

For this test, we restrict the factor set to a single asset class. There are two motivations to conduct this test:

- Testing the contribution of each asset class to the overall crisis prediction.
- To a lesser extent, it also allows us to test if the results that we presented are not conditioned to the choice of a specific dataset.

Here, our analysis is restricted to the asset classes that contains enough factors to estimate a distribution, Real Estate, Hedge Funds and Volatility thus being excluded.

Table 4.11 shows that none of the asset classes dominate *per se* the aggregated factor set. It proves that the diversity of factor asset classes contributes to the quality of the predictions.

According to these results, it seems clear that the Stock Markets is the most important asset class, which is a natural finding. However, most of the other asset classes also perform on this standalone basis, which is a strong sign that the trading strategy can perform independently of the choice of a particular factor set.

4.5.6 Sensitivity to Trading Strategy

We now discuss the extent to which the performances that we observe are the results of our particular choices of leverage for our trading strategy. Since the full sample volatility of the returns is standardized to make the market and the trading strategy comparable, what controls the aggressiveness of the strategy is the magnitude of the difference between the risk OFF leverage and the risk ON leverage, more than the level of the leverage itself. Hence, we try some changes of the risk OFF leverage, while keeping the risk ON leverage constantly equal to 2.

Moreover, we try an In/Out strategy, in which the leverage is set to 0 when the risk is OFF.

Table 4.12 Sensitivity of the average improvement of the Sharpe ratio to the leverage used

Risk OFF leverage I Risk ON leverage	Improvement of the average (%)	p-value (%)	Percentage of positive improvements (%)
Risk OFF −1 I Risk ON 2	201.88	0.14 (***)	91.67
Risk OFF −0.5 I Risk ON 2	215.39	0.09 (***)	100.00
Risk OFF 0 I Risk ON 2	201.83	0.07 (***)	100.00
Risk OFF 0.25 I Risk ON 2	182.78	0.06 (***)	100.00
Risk OFF 0.5 I Risk ON 2	157.49	0.06 (***)	100.00
Risk OFF 1 I Risk ON 2	99.19	0.08 (***)	100.00

Finally, we also allow for short sales, making the strategy extremely aggressive and dangerous, as the markets are known to go up on average.

Below are the results, for these different settings (Table 4.12):

Remarkably, the *p*-value is quite insensitive to the level of performance, which is a good indicator that the methodology we proposed for computing this *p*-value is reliable.

Note that the performances are always increasing when we increase aggressiveness, until the −1I2 case. It is an extremely strong sign of predictive power that the strategy survives to short sales, and even gets better under this framework. However, if one is going to use short sales, it should be done carefully.

Indeed, in the −0.5I2 case, the average random improvement is −49%, which clearly shows that shorting the market is an extremely dangerous strategy, for which we should have an extreme level of confidence in the underlying signal.

Still, if the strategy is successful, this is also to an extreme point: below, we display the example of the Stoxx Europe 600, for which the −0.5I2 strategy reaches a final level of 1,242.44 points (base 100 in 2001), versus a final level of 165.92 points for the market (recall that both portfolio returns have the same volatility) (Fig. 4.11).

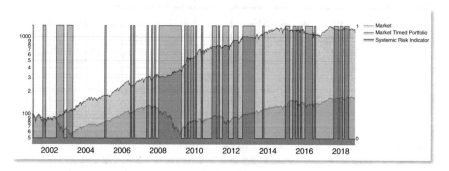

Fig. 4.11 Stoxx Europe 600: Market Timed Portfolio with aggressive leverage strategy, vs Market Portfolio (equal volatility)

4.6   Conclusions

In this chapter, we presented a polymodel-based Systemic Risk Indicator and its associated trading strategy.

The predictive power of the indicator was shown to be statistically and economically significant across several markets and time periods. These results are found to be robust in the tests that have been conducted. However, the robustness of the strategy, as well as its performance, can be increased by selecting/combining better/different parameters of the strategy.

In terms of improving the design of the indicator, it is possible to explore different paths. Subject to computational resources, estimating the indicator on a daily frequency can be a game changer since the indicator spikes sometimes only a few days after the beginning of a market drawdown. The methodology developed for the indicator should also be tested to see if it can predict market rallies. Indeed, it is possible that some particular herding behavior may occur at the beginning of speculative bubbles, for example. If this is the case, it can also improve the crisis indicator by differentiating exuberant mimicry from collective panic.

This last point emphasizes the need to strengthen the economical foundations of the indicator. It is especially important to better understand which factors correlate or de-correlate the most before the crisis, and why they behave accordingly.

The indicator should also be compared with existing alternatives, to measure how it can add value to a portfolio that aims at benefiting from various market timing signals (for example, its performance can be compared with other market stress indicators, such as Chicago Fed's National Financial Conditions Index (NFCI)).

There are already several systemic risk indicators identified by the literature. However, as far as we know, at the time of writing, all these indicators are based on the increase of correlations. The present chapter adds value to the literature by identifying another, simultaneous and possibly stronger phenomenon of de-correlation of the factors. Considering the shape of the representative pre-Crisis distributions and the results achieved by the Systemic Risk Indicator, this finding may be one of the keys to understanding the financial crisis.

References

Douady, R., & Kornprobst, A. (2018). An empirical approach to financial crisis indicators based on random matrices. *International Journal of Theoretical and Applied Finance, 21*(03), 1850022.

Geraskin, P., & Fantazzini, D. (2013). Everything you always wanted to know about log periodic power laws for bubble modelling but were afraid to ask. *The European Journal of Finance, 19*(5), 366–391.

Harmon, D., de Aguiar, M. A., Chinellato, D. D., Braha, D., Epstein, I., & Bar-Yam, Y. (2011). Predicting economic market crises using measures of collective panic. Available at SSRN 1829224.

Johansen, A., Ledoit, O., & Sornette, D. (2000). Crashes as critical points. *International Journal of Theoretical and Applied Finance, 3*, 219–255.

Junior, L. S., & Franca, I. D. P. (2012). Correlation of financial markets in times of crisis. *Physica A: Statistical Mechanics and its Applications, 391*(1–2), 187–208.

Parzen, E. (1962). On estimation of a probability density function and mode. *The Annals of Mathematical Statistics, 33*(3), 1065–1076.

Patro, D. K., Qi, M., & Sun, X. (2013). A simple indicator of systemic risk. *Journal of Financial Stability, 9*(1), 105–116.

Rosenblatt, M. (1956). Remarks on some nonparametric estimates of a density function. *The Annals of Mathematical Statistics, 27*, 832–837.

Silva, W., Kimura, H., & Sobreiro, V. A. (2017). An analysis of the literature on systemic financial risk: A survey. *Journal of Financial Stability, 28*, 91–114.

Silverman, B. W. (1986). *Density estimation for statistics and data analysis* (p. 45). Chapman & Hall/CRC.

Sornette, D. (2009). Dragon Kings, Black Swans and the prediction of crisis. *International Journal of Terraspace Science and Engineering, 2*(1), 1–18.

Ye, X., & Douady, R. (2019). Systemic risk indicators based on nonlinear polymodel. *Journal of Risk and Financial Management, 12*(1), 2.

Zheng, Z., Podobnik, B., Feng, L., & Li, B. (2012). Changes in cross-correlations as an indicator for systemic risk. *Scientific Reports, 2*, 888.

Chapter 5
Predictions of Industry Returns

Abstract We present a new factor predicting the cross-section of industry returns on the stock market. The factor is based on a measure of antifragility, pointing out that the non-linearity of the link between market and industry returns is priced by market participants. It is shown to differ from other well-known industry factors, including downside beta and coskewness factors, which are entirely subsumed by the antifragility factor. A trading strategy derived from the factor exhibits a Sharpe ratio of 1.10, and successfully resists various robustness tests.

Keywords Antifragility · LNLM modeling · Cross-Section of industry returns · Trading strategy · Trading signal · Stock returns predictions · Coskewness · Downside Beta

5.1   Introduction

In this chapter, we predict the returns of the industry component of the stock returns. More precisely, we investigate the presence of a market antifragility factor that predicts future industry returns.

Since Fama and French (1993), there has been a long tradition of identifying factors which explain the cross-section of stock returns. However, only a few papers tackle the problem of predicting the cross-section of industry returns, one of the most famous examples being Asness et al. (2000). In the current chapter, we thus add to the literature on this particular topic.

We measure market antifragility as the convexity of the non-linear response of the industry return to a market shock. The economic rationale that underlies the performance of the antifragility factor can be seen evenly from both the behavioral and classical financial approach of the efficient markets.

Seen from the efficient markets hypothesis, the antifragility factor can be interpreted as a risk premium. The portfolio built using the antifragility signal takes long positions on fragile industries and short positions on anti-fragile industries. Hence, for extreme market events, the portfolio suffers from its concave position on the industries, especially since its losses accelerate with the magnitude of the market move. Since such events occur with a low probability, the resulting behavior of the portfolio would be some punctual crashes. These drawdowns are clearly identifiable in the P&L curve of the simulation. Thus, the persistent abnormal returns of the fragile industries may be seen as a compensation for the extreme risk that they carry, i.e. as a risk premium.

The antifragility factor may also be justified from the behavioral finance point of view. Indeed, we may consider that fragile industries constitute an investment opportunity just because investors over-buy the antifragile industries. Such a behavior would be strongly consistent with a loss-aversion bias of investors who would prefer to buy anti-fragile industries at a very high (too high) price than bear the risk of enduring extreme losses.

The market risk as a factor which explains the cross-section of stock returns has already been widely explored.

The low beta anomaly, as presented for example in Baker et al. (2011), is a well-known predictor of the cross-section of stock returns. Low beta refers to the parameter estimated in a linear model with the stock returns as target variable and market returns as explanatory variable. In the long run, low-beta (and low-vol) stocks outperform the high-beta (high-vol) stocks. Such a phenomenon is more in line with a market anomaly than with a risk premia explanation, since the riskier stocks perform the worst. This is exactly the opposite of what we observe with our antifragility risk measure, which does not make the low beta anomaly a very promising candidate to explain it. Nevertheless, since the beta is the first derivative of a linear model while the antifragility is the average second derivative of a non-linear model, we should control for it.

The downside beta, as presented by Ang et al. (2006), is a risk measure designed to capture downside risk. It refers to a market beta estimated as:

$$\beta^- = \frac{cov\left(r_a, r_m | r_m < \bar{r}_m\right)}{var\left(r_m | r_m < \bar{r}_m\right)} \tag{5.1}$$

Here r_a is the return of a given stock a, r_m is the return of the market, and $\bar{r}_m$ is the average return of the market.

So, it is simply a beta evaluated while only taking into account the bearish returns of the market. Such a metric is by nature highly correlated with the simple market beta, which pushed Ang and colleagues to introduce a relative downside beta, which is the metric we focus on below:

$$\beta^{relative} = \beta^- - \beta. \tag{5.2}$$

In the paper, the relative downside beta is a risk premium: stocks having a high downside beta are associated with high future returns to compensate the risk of holding the stock. Such a behavior can be explained by loss aversion, and goes in the same direction as the antifragility anomaly.

The coskewness is another measure of the market risk of a given stock, more focused on extreme events, such as antifragility. It can be defined as:

$$coskew = \frac{E\left[(r_a - \bar{r}_a)(r_m - \bar{r}_m)^2\right]}{\sqrt{var\,(r_a)var\,(r_m)}}. \tag{5.3}$$

Harvey and Siddique (2000) already proposed coskewness as a risk premium. We thus control that we are not proposing a variation of this risk factor.

Although all these anomalies have essentially been explored to predict the cross-section of stock returns, the underlying rationale seems applicable to explain the cross-section of industry returns. We thus benchmark and stress the antifragility anomaly with these other predictors in order to control for the reality of our contribution to the literature. Indeed, the antifragility anomaly seems to be stronger than any other anomaly that has previously been identified in the literature, at least in relation to the particular question of the industry returns.

The chapter is organized as follows:

- We first present the data used to conduct our empirical analysis.
- We then define the methodology used to measure antifragility and to construct a portfolio which takes industry bets on the cross-section of stock returns.
- Empirical results are presented, including the performance of a trading strategy which benefits from the antifragility factor.
- Several robustness tests are performed to control for the reliability of the results, after which we conclude the chapter.

5.2   Data

We use data from Refinitiv (ex-Reuters) to get daily returns, market capitalizations and industry classification as defined by the GICS 3 classification (Global Industry Classification Standard, MSCI), all measured using end-of-day data in USD. Our sample covers a period from 1999-01-01 to 2018-09-30.

Our investment universe is composed of the 500 largest market capitalizations[1] in the US stock markets, restated monthly.

"Industries" is defined as equally-weighted buckets of stocks belonging to the same industry. The "Market" is an equally-weighted portfolio of industries.

---

[1] In a concern for realism, all the stocks have been pre-filtered to reach a median monthly volume of $1 million in the month preceding their selection for the investment universe.

Table 5.1 Summary statistics of industry buckets: Counts of sub-elements (stocks), aggregated statistics

Mean	Std	Min	Median	Max
8.10	7.07	1.00	6.00	52.00

Table 5.2 Summary statistics of industry buckets: Counts of sub-elements (stocks), aggregated quantiles

0.00	0.05	0.10	0.25	0.50	0.75	0.90	0.95	1.00
1.00	1.00	1.00	3.00	6.00	12.00	17.00	21.00	52.00

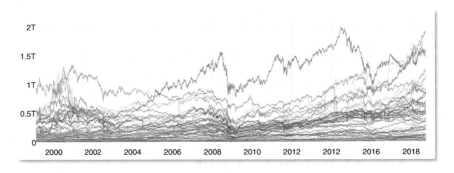

Fig. 5.1 Market capitalizations of industry buckets over time

5.2.1 Summary Statistics

The use of the GICS 3 classification leads to a universe of 66 Industries.[2] A simple count of elements inside each bucket reveals large disparities that can occur among the different sectors over time (Table 5.1):

The table of quantiles (Table 5.2) on the aggregated counts of elements over time shows that there are a few industries that incorporate a large number of elements:

We clearly observe a Pareto-like distribution of the number of elements, which motivates the use of an equal-weight methodology for the definition of the market portfolio. Indeed, we would like to capture the reaction of industry to its economic environment, approximated by a representative market, and not only by a few industries. This thinking is further supported by the plot of the industries market capitalizations over time (Fig. 5.1), and by their table of quantiles (Table 5.3).

Appendix: Industry Buckets Summary Statistics describes the statistics in detail, industry per industry.

---

[2]GICS Levels are named Sectors for level 1, Sub-Sectors for level 2, Industries for level 3, and Sub-Industries for level 4. We refer indistinctly to "industries" or "sectors" in the chapter for level 3 of the GICS classification.

Table 5.3 Market capitalizations of industries

	0.00	0.05	0.10	0.25	0.50	0.75	0.90	0.95	1.00
Average market capitalization quantiles	8.86E +09	1.21E +10	1.60E +10	3.74E +10	1.25E +11	3.16E +11	4.74E +11	6.32E +11	1.05E +12

5.3 Methodology

5.3.1 Measuring Antifragility

Antifragility was introduced by Taleb (2012), and its mathematical aspects were detailed by Taleb and Douady (2013). Taleb (2013) offers the following definition of the concept of antifragility: "Simply, antifragility is defined as a convex response to a stressor or source of harm (for some range of variation), leading to a positive sensitivity to increase in volatility (or variability, stress, dispersion of outcomes, or uncertainty, what is grouped under the designation 'disorder cluster'). Likewise fragility is defined as a concave sensitivity to stressors, leading to a negative sensitivity to increase in volatility."

At first, one may consider this concept to be close to robustness or resiliency. Taleb (2012) clarifies this point as follows: "Antifragility is beyond resilience or robustness. The resilient resists shocks and stays the same; the antifragile gets better.".

Such a property seems desirable for equities, since any investor would like the components of their portfolio to over-perform the market in bad circumstances.

Generally, measuring the antifragility of a variable Y via its sensitivity to a variable X can be done in two simple steps:

- Estimate a non-linear function of reaction $\varphi(X)$ between X and Y, for example using the Linear-Non-Linear-Mixed model.
- Compute the AntiFragility Indicator as the integral of the second derivative of the function of reaction on the domain[3] of X. We approximate the function by a sum in which the X_i's are sampled at regular intervals on the domain of X (such an approximation may be improved by a non-parametric estimation of $p(X)$):

$$AFI = \int_{X_{min}}^{X_{max}} \varphi''(X) \times p(X) \ d(X) \cong \sum_{c=X_{min}}^{X_{max}} \varphi''(X_c). \tag{5.4}$$

---

[3] By default $[X_{min}: X_{max}] = [-10\%: +10\%]$ thereafter.

5.3.2 Predicting Industry Returns

The intuition behind the following predictions is that antifragility can be a part of the large literature dedicated to factor investing. Indeed, this literature reveals that the cross-section of stock returns is highly multidimensional. The interested reader may refer to Green et al. (2014) and Hou et al. (2017) for an exhaustive review of the existing factors. Again, several publications point out that factor investing is also possible among industries, by taking long positions on some industries and short on others to form a portfolio (e.g. Asness (2000)).

Following the previous studies, we assess the predictive power of their antifragility levels on future industry returns.

We measure the antifragility using the AntiFragility Indicator (AFI) as defined in Section 5.3.1. A key question in order to describe antifragility in a particular context is to define what the object of analysis is sensitive to. The market is a convenient and natural choice, since it may be the first vector of shocks that interact with industries. Even if we only have access to daily data, we try to identify low-frequency response of industries, using 21-day cumulative returns re-computed every day for the LNLM estimations. This allows us to see how the industries react over the month during bearish and bullish markets. For the question of robustness, we use a very large rolling window of 5 years, allowing 1,260 points for each fit (monthly returns are recomputed on a daily basis). For computational reasons, we only re-compute the AFI every 3 months.

To represent the market in the most faithful manner, once a stock enters the investment universe, we keep it in the industry bucket used to compute AFI even after it exits the investment universe. This allows us to get more representative industry returns, even if we then only partially bet on them, since we restrict our positions to the investment universe.

Once the AFI is computed, we assess its predictive power using several means.

In a first step, we implement a Long/Short trading strategy.

The inputs of the Long/Short portfolio are a normalized version of the base AFI signal. At each date, we sort the industries by their AFI in descending order, and we normalize the ranked signal obtained to be uniformly distributed between -1 and 1. Indeed, we are only interested in measuring the impact of the relative industry antifragility on future returns, and having an average of 0 for the signal allows us to partly neutralize the market as a source of portfolio returns. Such a procedure of ranking and normalization is straightforward in factor investing strategies (see for example Chan et al., 1998, or Coqueret and Guida, 2018). For convenience, in the following we use the initialism "AFI" to refer to the normalized AntiFragility Indicator.

The signal is then mapped to the sub-components of the industries, all the stocks belonging to an industry bucket sharing the same industry score. This operation is done for the sake of realism, as there is no real-world vehicle to invest directly in our industry buckets.

The portfolio weights are computed by optimizing a characteristic portfolio in order to reflect the score mapped on stocks. The choice of the characteristic portfolio is consistent with our concern to obtain a portfolio that reflects the predictive power of the antifragility characteristic. For a given vector of attributes (or characteristics) of assets $\kappa = [\kappa_1, \kappa_2, \ldots, \kappa_b]$, and for portfolio weights on assets $w = [w_1, w_2, \ldots, w_b]$, the exposure of the portfolio to the given characteristic is simply:

$$\sum_{a=1}^{b} \kappa_a w_a. \tag{5.5}$$

The characteristic portfolio of attribute κ is defined as the portfolio that has minimum risk (in terms of variance) and unit exposure to the characteristic κ (i.e. Eq. (5.5) is equal to 1). For example, the characteristic portfolio of the assets' betas is the market portfolio. The holdings of the portfolio respecting these properties can be obtained by a simple closed formula[4] (Grinold & Kahn, 2000):

$$w_k = \frac{\sum^{-1} k}{k^T \sum^{-1} k}. \tag{5.6}$$

Here Σ is the variance − covariance matrix of assets.

Thus, the characteristic portfolio is a convenient choice to perform a mean variance optimization while reflecting well the predictive power of a given characteristic. The resulting holdings are standardized to have an ex-ante target volatility of 10%.

In a second step, we perform a more theoretical exercise, and regress the 3-month future returns on the AFI.

5.4 Results

The LNLM fits show the interest in using non-linear modeling. As an example, in Fig. 5.2 below we display the fits for all industries on 2008-03-10:

Whatever the industries, Figure 5.2 shows that none of them adopt a completely linear behavior.

5.4.1 Long/Short Trading Strategy

We compute the P&L of the strategy every day using optimized weights computed 2 days before. This 2-day lag is motivated by the following assumed trading process: at date t, after the close of the US market, we collect all the needed data, compute the

---

[4]Since we are at the stock level and we don't want to penalize intra-industry risk, the variance-covariance matrix is approximated by its diagonal in the optimization process.

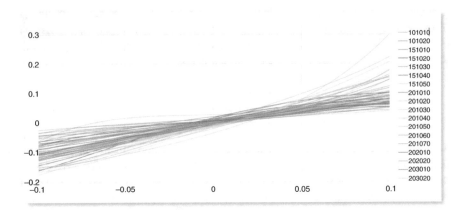

Fig. 5.2 LNLM modeling of industry returns by market returns

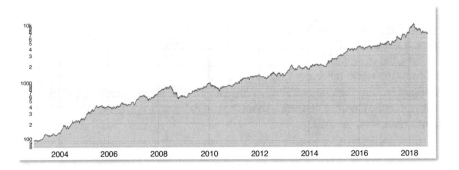

Fig. 5.3 P&L of the antifragility factor strategy

AFI and optimize a characteristic portfolio. The resulting trades that must be done to adjust the position are sent to the market in the morning of date t + 1 and executed during the day so that the optimized positions are reached at the close of t + 1. So, at date t + 2, the P&L are the returns on the position reached at the end of the previous day, i.e. the optimized weights of date t.

We present below the P&L in the form of a cumulative (compounded) P&L curve, displayed in log-scale, as it allows us to easily compare different periods, and thus evaluate the stability of the profitability of the strategy over time (Fig. 5.3). We start from a base 100 in 2004. Over the full period, the Sharpe ratio[5] of the L/S strategy is 1.10.

The above picture clearly indicates that over a sufficiently long period, the strategy is profitable. Here are the strategy's main statistics (Table 5.4):[6]

---

[5] Sharpe ratios are computed using the risk-free rate provided by Kenneth French's website.

[6] Turnover is defined as the number of times the unleveraged portfolio rotates each year.

Table 5.4 Performance statistics of the antifragility trading strategy

Annual excess return	Sharpe ratio	Turnover	Annual volatility	Worst drawdown 5 days	Worst drawdown 21 days	Worst drawdown 252 days	Worst drawdown Full Period
29.23%	1.10	5.73	26.58%	−16.37%	−23.31%	−28.51%	−42.01%

Table 5.5 Regression of the AFI on future returns

	Coef	Std err	t-stat	$P > |t|$	[95.0% Conf. Int.]	
Const	0.0005	0.0000	140.307	0.00E+00	0.001	0.001
AFI	0.0317	0.006	5.032	0.00E+00	0.019	0.044

Note that the volatility control applied ex-ante on the positions is clearly insufficient to reach an acceptable level of volatility (the initial target was 10%, which is far below the realized volatility of 26%). Still, this is of weak impact on the Sharpe ratio, which is our main performance metric to assess the predictive power of the strategy.

The strategy suffers from large drawdowns in 2008 and 2018. Several explanations are possible for this phenomenon. Among them, we can surmise that the antifragility factor suffers from the subprime crisis and the 2018 quant crisis.[7] Indeed, such a pattern of drawdowns is in line with the exposure to extreme market moves that the AFI portfolio takes.

5.4.2   Regressions on Future Returns

We regress the 3-month cumulative future returns (without compounding) of the industries on the current values of the AFI to give a second assessment of its predictive power. We display below the results of this linear regression (Table 5.5):

The constant is of course extremely statistically significant, since it captures the returns of the market (recall that the average of the cross-sectional AFI score is 0). Still, the signal shows a high level of statistical significance, with a t-stat of 5.032, and a p-value of 4.84E$-$5%.

5.4.3   Comparisons to Classical Factors

A very important test in the quest for new factors consists of checking if the antifragility factor is able to add value on top of classical factors. In order to do so, we compute the P&L of characteristic portfolios of common factors, using exactly the same methodology as for the L/S strategy, and use it to attempt to explain the returns of the antifragility portfolio.

We of course primarily control for the market factor, defined as the returns of our market benchmark.

---

[7] Several unsuccessful tests have been performed to try to explain the drawdowns by signal fluctuations, including controls for autocorrelation, goodness of fits of LNLM estimations, historical dispersion and internal consistency of the signal. This is outside the scope of the present chapter, and should be completed by further research.

Table 5.6 Regression of the AFI returns on the CAPM returns

	Coef	Std err	t-stat	$P > |t|$	[95.0% Conf. Int.]
Const	0.0004	0.0001	3.939	0.00E+00	0.000 0.001
Equal-weights market portfolio	0.2158	0.015	13.936	0.00E+00	0.185 0.246

Table 5.7 Regression of the AFI returns on the Fama–French 3 factors returns

	Coef	Std err	t-stat	$P > |t|$	[95.0% Conf. Int.]
Const	0.0004	9.45E−05	3.91	0.00E+00	0.000 0.001
Equal-weights market portfolio	0.0007	0.021	0.032	9.74E−01	−0.040 0.041
Size portfolio	−0.1862	0.016	−11.617	0.00E+00	−0.373
Value portfolio	0.2286	0.02	11.562	0.00E+00	0.190 0.267

In a second regression, we integrate the classical value and size factors, in order to control for the standard factors of Fama and French (1993).[8] The value is computed as the book to market ratio defined as net assets (assets minus liabilities) divided by current market capitalization. We aggregate the stock level signal simply by averaging it inside of each industry. The size signal is minus the log of the market cap. We aggregate by taking the sum of the stock level signals.

In a third regression, we include the market portfolio returns, as well as a Low Beta portfolio, a Momentum portfolio, a Downside Beta portfolio, and a Coskewness portfolio. The motivation to add these statistical factors is that we suspect that they are linked to the antifragility factor, since they are computed using the same data, namely the returns. This reasoning is especially appropriate for Low Beta, since a very curvy shape of our fits may be approximated by a steep line. As explained in the introduction of the chapter, Downside Beta and Coskewness are also particularly relevant candidates to explain the returns of the antifragility anomaly. The Momentum signal is computed following the classical methodology of computing the 11-month return lagged by 1 month (Carhart (1997), see also Jegadeesh and Titman (1993)). We directly use the returns of our industry buckets here. The Low Beta is computed as the parameter of a simple univariate regression of the daily industry returns on the market (including a constant), using a 1 year rolling window. Downside Beta is computed using the formula presented in the introduction (to be perfectly precise, we continue to consider the relative downside beta), as well as Coskewness.

Below are the results of the three regressions. We standardized the volatilities of the P&Ls to be equal to 10% annual, to allow comparisons among coefficients (Tables 5.6, 5.7, and 5.8).

---

[8] It would have been natural to use the industry portfolios factor returns provided by Kenneth French for this purpose. However, industries are defined differently in French's data, so we have directly re-computed the portfolio performances in our framework in order to avoid the introduction of a methodological bias.

Table 5.8 Regression of the AFI returns on 5 statistical factors returns

	Coef	Std err	t-stat	P > ltl	[95.0% Conf. Int.]	
Const	0.0003	8.36E−05	3.142	2.00E−03	9.88e−05	0.000
Low Beta Portfolio	0.2015	0.015	13.575	0.00E+00	0.172	0.231
Equal-weights market portfolio	−0.0583	0.015	−3.803	0.00E+00	−0.116	
Momentum portfolio	0.0984	0.015	6.568	0.00E+00	0.069	0.128
Coskewness portfolio	0.2396	0.016	15.22	0.00E+00	0.209	0.270
Downside beta portfolio	0.3165	0.016	20.126	0.00E+00	0.286	0.347

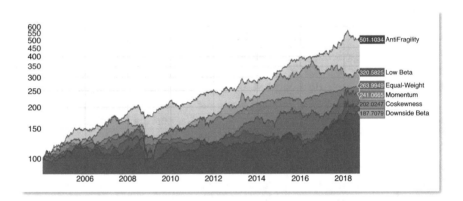

Fig. 5.4 Factor portfolios compounded returns (equal rolling volatility of 10% annual)

Whatever the factors used for the regression, we always keep a highly statistically significant constant below 0.2% of p-value. The constant is also economically significant, and corresponds to an abnormal annual return of around 17% in the last regression (without compounding, without volatility standardization before regression). Such a level of abnormal returns depends of course on the leverage and the volatility. The R^2 of the third regression being 30.3%, and the Sharpe ratio of the strategy being 1.10, we may expect the abnormal returns to be around 11% * $(1 - 0.303) = 7.67\%$ for a 10% realized volatility, which stays economically significant.

Below we show the P&L of the statistical portfolios, which seems to indicate that all 5 factors are performing at the industry level:

From Fig. 5.4, antifragility seems to be a relatively strong factor.

We can see from the correlation matrix of the statistical factors' P&L that antifragility is, as expected, particularly correlated with the Downside Beta and the Coskewness anomalies, which are risk premia that share the same rationale with antifragility (Fig. 5.5):

Such a level of correlation may indicate that Coskewness and Downside Beta capture the same source of alpha as antifragility. In order to better understand better

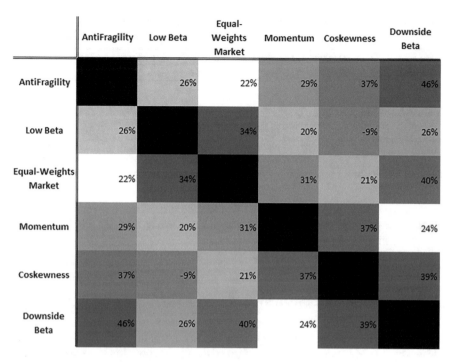

Fig. 5.5 Correlation matrix of factors returns

Table 5.9 Regression of the AFI returns on the Coskewness returns

	Coef	Std err	t-stat	$P > |t|$	[95.0% Conf. Int.]
Const	0.0004	9.29E−05	4.105	0.00E+00	0.000 0.001
Coskewness	0.3697	0.015	25.087	0.00E+00	0.341 0.399

Table 5.10 Regression of the Coskewness returns on the AFI returns

	Coef	Std err	t-stat	$P > |t|$	[95.0% Conf. Int.]
Const	0.0000287	9.31E−05	0.308	7.58E−01	−0.000 0.000
AntiFragility	0.3697	0.015	25.087	0.00E+00	0.341 0.399

the link among these 2 anomalies and Antifragility, we perform univariate linear regressions, in which we model Antifragility as a function of another anomaly, and inversely, in which we model the anomaly as a function of Antifragility. In the case of Coskewness, the results of the two estimated models are (Tables 5.9 and 5.10):

Hence, when trying to explain antifragility by coskewness, we keep a significant alpha (R^2 is 13.7%), while the reverse is not true. Such a result tends to indicate that antifragility subsumes the Coskewness anomaly at the industry level.

We obtain similar results for the Downside Beta anomaly (R^2 is 21.4%) (Tables 5.11 and 5.12):

Table 5.11 Regression of the AFI returns on the Downside Beta returns

	Coef	Std err	t-stat	$P > |t|$	[95.0% Conf. Int.]	
Const	0.0004	8.86E−05	4.028	0.00E+00	0.000	0.001
Downside Beta	0.4623	0.014	32.866	0.00E+00	0.435	0.490

Table 5.12 Regression of the Downside Beta returns on the AFI returns

	Coef	Std err	t-stat	$P > |t|$	[95.0% Conf. Int.]	
Const	−2.551E−07	8.88E−05	−0.003	9.98E−01	−0.000	0.000
AntiFragility	0.4623	0.014	32.866	0.00E+00	0.435	0.490

These results would thus imply that the extreme market risk that industries bear is better captured and priced by the antifragility anomaly than by previous signals identified in the literature.

5.4.4 Separating Concavity and Convexity

Separating concavity and convexity allows us to control that the returns are not just driven by one of these two characteristics of the impact function. Convexity reflects the antifragility of the industries, i.e. their ability to perform extremely well when there is an extreme market deviation. Concavity, on the other hand, truly captures their extreme risk, as a non-linear concave response would lead to dramatic losses. Hence controlling for concavity and convexity effects is equivalent to controlling for the importance of the propensity of the industries to lose and for their propensity to win.

Our simple test to differentiate both is to only keep positive or negative second derivatives in the computation of the AFI. We then re-apply our test using the L/S strategy.

Below are the summary tables (Table 5.13) and portfolio P&L curve (Fig. 5.6), for the convexity indicator first:

And here are the results for the concavity indicator (Fig. 5.7; Table 5.14):

Although there is a slightly higher Sharpe for convexity, it seems clear that both positive and negative derivatives matter, leading to a more stable (and higher) profitability over time.

5.4.5 Separating Positive and Negative Market Returns

It may make sense to consider that investors take more care about antifragility when the market shows an extreme loss. To control for this effect, we compute the antifragility signal using the second derivatives only when the market returns are

Table 5.13 Performance statistics of the antifragility trading strategy (convexity only)

Annual excess return	Sharpe ratio	Turnover	Annual volatility	Worst drawdown 5 days	Worst drawdown 21 days	Worst drawdown 252 days	Worst drawdown full period
22.42%	1.00	5.98	22.52%	−11.03%	−24.42%	−33.44%	−38.31%

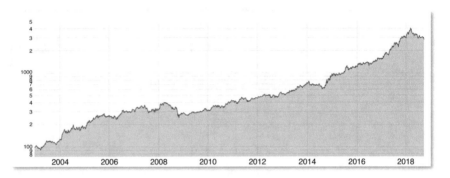

Fig. 5.6 P&L of the antifragility factor strategy (convexity only)

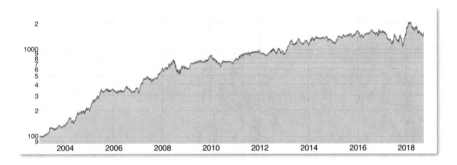

Fig. 5.7 P&L of the antifragility factor strategy (concavity only)

positive, and only when the market returns are negative. Below are the results of this split, for the positive returns only (Fig. 5.8; Table 5.15):

Below are the results recomputed with negative market returns only (Fig. 5.9; Table 5.16):

This suggests that the investors care more about antifragility in the case of positive market returns. Still, both versions of the signal continue to deliver significant returns, and none of them outperform the original version. Antifragility thus appears to be a broad phenomenon, priced by the market both during negative and positive market moves, as well as through concavity and convexity.

5.5 Robustness Tests

This section is dedicated to performing various tests to assess the robustness of our results, with regard to several considerations.

Table 5.14 Performance statistics of the antifragility trading strategy (concavity only)

Annual excess return	Sharpe ratio	Turnover	Annual volatility	Worst drawdown 5 days	Worst drawdown 21 days	Worst drawdown 252 days	Worst drawdown full period
18.91%	0.78	5.96	24.39%	−11.51%	−24.60%	−31.52%	−37.99%

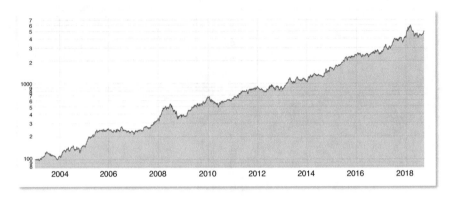

Fig. 5.8 P&L of the antifragility factor strategy (positive market returns only)

5.5.1   Stability of the Signal

We first check the stability of the performance regarding the speed of trading. The turnover, of course, looks relatively low to the practitioner, but we would like to confirm this point by computing the performance of various lags of the portfolio's positions. By showing what would be the Sharpe ratio of the strategy if we would have reached the optimal positions 3 days, 5 days, 10 days, etc. after obtaining the optimized weights of the portfolio, we assess the profitability of the strategy at low trading frequencies. The Table 5.17 summarizes our findings:

The strategy clearly shows a high level of stability, since even an investor trading a quarter after its optimization would make profits. This is not too surprising since the AntiFragility Indicator is very stable because of the 5-year rolling window used for its computation. It moderates the concerns we may have about the impact of re-computing the AFI only every 63 days. This point is further confirmed by the high daily autocorrelation of the score, which is 99.84%. The autocorrelation of the current score to its 21-day lag is thus still 97.07%.

Investment strategies usually exhibit a link between the stability of the factor on which they are based and the stability of the performance. While this stability may indicate that the strategy is easily tradable, it is also associated with a risk of "cherry picking" of the industries at the beginning of the trading period. Indeed, we may randomly create a portfolio that simply picks the good industries at the beginning of the simulation and holds them for the duration of the full simulation, while this is clearly less likely to happen if the sign (long/short) of the holdings changes frequently over time. Below we give the summary statistics of the number of sign changes of the holdings (Table 5.18):

An average number of 6 changes over the 14-year simulations between long and short positions can be interpreted as a reasonable protection against a lucky pick. Note that, with no exceptions, every industry changes its holding sign at least once during the full simulation. The time series of each of the industry scores are available in the Appendix: AFI Scores per Industry over time.

Table 5.15 Performance statistics of the antifragility trading strategy (positive market returns only)

Annual excess return	Sharpe ratio	Turnover	Annual volatility	Worst drawdown 5 days	Worst drawdown 21 days	Worst drawdown 252 days	Worst drawdown full period
26.47%	1.06	5.81	24.93%	−14.87%	−20.40%	−17.13%	−37.80%

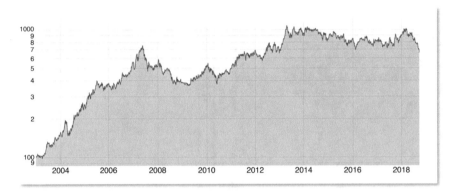

Fig. 5.9 P&L of the antifragility factor strategy (negative market returns only)

5.5.2 Sensitivity to Hyper-Parameters

We also control for the resistance of the signal if computed with different hyper-parameters. Two hyper-parameters seem to be of real importance in the computation of the AntiFragility Indicator: the temporal depth of the rolling window, and the domain on which the derivatives are evaluated (X_{Min} and X_{Max} in the AFI definition).

We first re-compute the performance for different rolling windows (Table 5.19):

Clearly the window has an influence on the performance, but all the values tested continue to deliver a significant performance. The differences observed are explained by the quantity of observations being too low to accurately model the reaction function for small windows, and the observations being too old and no longer relevant for the fit, concerning the large windows.

We then assess the sensitivity of the performance to the domain of the derivatives (Table 5.20).

Here, while we can still observe a sensitivity of the Sharpe ratio to the hyper-parameter, we find again a strong robustness of the signal performance. Very interestingly, the signal computed with derivatives between +1% and −1% still deliver significant performance, despite the fact that this domain is very small, considering that we are working with 21-day cumulative returns. This point highlights the importance of the choice we made in the definition of the AFI. Since we approximated the probability weighted integral by a sum of *regularly* spaced derivatives, we indirectly assumed a uniform distribution for the associated probabilities. The tightening of the domains of derivatives thus shows that the central derivatives matter a lot for the performance, while the extremes may be less important. Assuming a bell distribution for the market returns, one may understand this as a reflection of the probabilities of the derivatives: what occurs less often simply matters less. Further research may try to weight the approximated AFI by the probability distribution of market returns.

Table 5.16 Performance statistics of the antifragility trading strategy (negative market returns only)

Annual excess return	Sharpe ratio	Turnover	Annual volatility	Worst drawdown 5 days	Worst drawdown 21 days	Worst drawdown 252 days	Worst drawdown full period
14.18%	0.54	5.73	26.25%	−12.89%	−22.40%	−35.48%	−51.53%

Table 5.17 Sharpe ratios for different lags of the optimized weights

Lag 2 days	Lag 3 days	Lag 4 days	Lag 5 days	Lag 10 days	Lag 15 days	Lag 21 days	Lag 42 days	Lag 63 days
1.10	1.09	1.07	1.05	1.04	1.05	1.07	1.07	0.98

Table 5.18 Summary statistics: Number of changes of position direction (L/S)

Mean	Std	Min	Max	Median
6.62	4.09	1.00	24.00	6.00

Table 5.19 Performance statistics for different rolling windows

	Annual return (%)	Annual volatility (%)	Sharpe ratio
3 years	14.55	26.80	0.49
4 years	29.71	27.06	1.05
5 years	30.41	26.58	1.10
6 years	26.87	26.75	0.96
7 years	24.52	25.80	0.90

Table 5.20 Performance statistics for different domains of derivatives

	Annual return (%)	Annual volatility (%)	Sharpe Ratio
[Xmin: Xmax] = [−1%: +1%]	22.51	24.93	0.86
[Xmin: Xmax] = [−5%: +5%]	26.04	25.56	0.97
[Xmin: Xmax] = [−10%: +10%]	30.41	26.58	1.10
[Xmin: Xmax] = [−15%: +15%]	24.95	25.92	0.92
[Xmin: Xmax] = [−20%: +20%]	19.42	25.53	0.71

Table 5.21 Performance statistics for different seeds

	Min	Mean	Median	Max	Std
Annual return	30.13%	30.60%	30.56%	31.63%	0.30%
Annual volatility	26.37%	26.51%	26.51%	26.64%	0.06%
Sharpe Ratio	1.09	1.11	1.11	1.15	0.01

In order to account for the randomness of the fits done using the LNLM model, we seeded the generator that randomly selects the observations which belong to each fold in the computation of μ. This practice makes it possible to replicate the results. If we used the default seed,[9] we still need to control that the choice of a particular set of random draws hadn't led to a spuriously high level of performance.

Hence, we launched L/S simulations for 50 different seeds,[10] and present below the resulting performance statistics. Overall, it clearly shows that the simulations are very weakly sensitive to the choice of a particular seed (Table 5.21).

---

[9] Seed #0 in the numpy library for python.

[10] Numbered from 0 to 49.

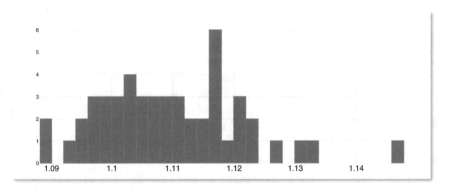

Fig. 5.10 Distribution of the Sharpe ratios for different seeds

This finding was further confirmed by considering the volatility of the distribution of the Sharpe ratios achieved (Fig. 5.10):

Eventually, we computed the correlations between the scores and the optimized weights, at the stock level and at the industry level.[11] This last test is performed as a sanity check that indicates if the outputs of the characteristic portfolio methodology really reflect its inputs. The Pearson correlation between the AFI and optimized weights is 81.17% at the stock level and 87.13% at the industry level. We also computed the Spearman correlation, as in our case the ranking is more important than the precise levels (since the optimization process minimizes the variance, we may expect slightly different weights for a same level of score). The Spearman correlation between the AFI and optimized weights is 91.54% at the stock level and 93.14% at the industry level.

5.6 Conclusions

The present chapter proposes an original approach to predict the industry returns on the stock market. Once again, it emphasizes the fact that the multi-dimensionality of the cross-section of stock returns is still a wide territory to explore for the literature. The emergence of alternative data is a promising path for this journey, but it should not mask the need for an improvement to the modeling methodologies.

In particular, using very common data (stock returns only), the present chapter demonstrates that the standard linear modeling approaches would have failed to capture the subtlety of the reality. Non-linear modeling matters, and its utility is not restricted to tail modeling, since antifragility is also priced in the center of the distribution. This application of the LNLM model also gives insights about its

---

[11] Industry level optimized weights are obtained by a simple averaging of the weights that belongs to the same industry.

efficiency and ability to reflect true relations between target and explanatory variables.

A lot of research questions emerge from this chapter. Considering the tests performed, our results seem robust, and show the high importance of the antifragility factor to explain the cross-section of industry returns. This level of importance in the prediction of future returns should be confirmed using other datasets and/or alternative antifragility measures. The computation of the AFI itself should be improved by the weighting of the second derivatives' probabilities.

The regressions on the P&L of other factors also showed statistically significant links between the antifragility factor and well-known factors. Their interactions should be studied, evenly from an economics or statistical point of view.

Finally, antifragility subsumes Downside Beta and Coskewness in our study. This finding should be studied directly in the cross-section of stock returns, for which it can significantly improve the existing knowledge on risk premia.

References

Ang, A., Chen, J., & Xing, Y. (2006). Downside risk. *The Review of Financial Studies, 19*(4), 1191–1239.

Asness, C. S., Porter, R. B., & Stevens, R. L. (2000). *Predicting stock returns using industry-relative firm characteristics.* Available at SSRN 213872.

Baker, M., Bradley, B., & Wurgler, J. (2011). Benchmarks as limits to arbitrage: Understanding the low-volatility anomaly. *Financial Analysts Journal, 67*(1), 40–54.

Carhart, M. M. (1997). On persistence in mutual fund performance. *The Journal of finance, 52*(1), 57–82.

Chan, L. K., Karceski, J., & Lakonishok, J. (1998). The risk and return from factors. *Journal of Financial and Quantitative Analysis, 33*(2), 159–188.

Coqueret, G., & Guida, T. (2018). Stock returns and the cross-section of characteristics: A tree-based approach. *SSRN Electronic Journal.* 3169773.

Fama, E. F., & French, K. R. (1993). Common risk factors in the returns on stocks and bonds. *Journal of Financial Economics, 33*(1), 3–56.

Green, J., Hand, J. R., & Zhang, X. F. (2014). The remarkable multidimensionality in the cross-section of expected US stock returns. Available at SSRN, 2262374.

Grinold, R. C., & Kahn, R. N. (2000). *Active Portfolio Management: A Quantitative Approach for Producing Superior Returns and Controlling Risk.* McGraw-Hill.

Harvey, C. R., & Siddique, A. (2000). Conditional skewness in asset pricing tests. *The Journal of Finance, 55*(3), 1263–1295.

Hou, K., Xue, C., & Zhang, L. (2017). *Replicating anomalies* (No. w23394). National Bureau of Economic Research.

Jegadeesh, N., & Titman, S. (1993). Returns to buying winners and selling losers: Implications for stock market efficiency. *The Journal of Finance, 48*(1), 65–91.

Taleb, N. N. (2012). *Antifragile: Things that gain from disorder.* Random House Trade Paperbacks.

Taleb, N. N. (2013). Philosophy: 'Antifragility' as a mathematical idea. *Nature, 494*(7438), 430–430.

Taleb, N. N., & Douady, R. (2013). Mathematical definition, mapping, and detection of (anti) fragility. *Quantitative Finance, 13*(11), 1677–1689.

Chapter 6
Predictions of Specific Returns

Abstract For each firm in the cross-section of stock returns we construct a prediction of its future returns. Each prediction is made with a dedicated polymodel estimated using a large set of explanatory variables, capturing the entire economic environment of each stock. A trading strategy is built on the predictions in order to assess their quality, reaching a Sharpe ratio of 0.91, being different than classical factors, and resisting a large set of robustness tests. Through the implementation of the trading strategy, we propose a method to tackle the problem of the aggregation of the predictions of a polymodel, based on the information added by each elementary model.

Keywords Polymodel theory · Artificial intelligence · Machine learning · LNLM modeling · Cross-section of stock returns · Trading strategy · Trading signal · Stock returns predictions

6.1 Introduction

The usefulness of breaking down the stock returns into a market component and an industry component has been known since King (1966), and we have known since Roll (1988) that the specific component of the returns, i.e. the residual movements of prices that do not depend on the market or industries, is even more significant than the other two factors. After focusing on providing predictions of the market returns in Chap. 4 and predictions of the industry returns in Chap. 5, we finally propose in the current chapter some predictions of the specific component of stock returns.

In order to do so, we perform one of the most straightforward, but challenging applications of Polymodel Theory: directly using the predictions of the target variable. Such an approach is challenging for at least two reasons: the selection of the predictors and the aggregation of the different predictions (see Chap. 2 for an in-depth, high-level discussion on these topics). We address and propose solutions to these two problems in the current chapter.

Showing that we can successfully predict the specific component of the returns of a single stock would lack generality. Hence, we provide predictions of the specific returns of 500 different stocks over a period of 15 years. These predictions are obtained by estimating dynamically (using a rolling window) a polymodel for each of the stock returns:

$$\{Y_a = \varphi_{i,a}(X_i) \quad \forall i\} \qquad \forall a \in [1:b].$$ (6.1)

Here a is the index of the stock and i is the index of the predictor.

All the elementary models are estimated using the LNLM model:

$$LNLM_a(X_i) \overset{\text{def}}{=} \bar{y}_a + \mu \sum_{h=1}^{u} \widehat{\beta}_{h,a,i}^{NonLin} H_h(X_i) + (1 - \mu)\widehat{\beta}_{a,i}^{Lin} X_i + \varepsilon_i.$$ (6.2)

We use a large set of predictors that includes 1,134 variables. One polymodel of 1,134 factors, re-estimated monthly for 15 years, represents around 200,000 elementary models to fit. Since the polymodels are fitted dynamically on 500 different target variables, this would result in fitting 102 million elementary models, which, due to 10-fold cross-validation, means more than 2 billion ordinary least squares estimations. This extremely large computation has of course been sped up using programming and mathematical tricks. For example, replacing the "Y" vector of the target variable in the OLS estimator formula by a matrix of *all* the target variables allows us to divide the number of matrix multiplications we need to perform by 500. On the I.T. side, an extensive use of multiprocessing allowed an important reduction of the computation time. Despite all these means of decreasing the computation time, this ambitious use of polymodels reduced our ability to perform a large number of tests, and hence made it difficult to maximize the quality of the final results.

The polymodel estimations should not be done using raw returns as target variables. Since we assume that the stock returns may be divided into a market, an industry and a specific component, we should extract from the raw returns the specific component. This is done using the residuals of a cross-sectional regression[1] with the industries used as categorical variables. This framework removes the impact of industries on the returns, but also implicitly the market, since the constant of the regression is the return of an equally weighted market portfolio. Compared to other possibilities (panel, time-series regressions), the cross-sectional regressions have the advantage of being instantaneous.

We estimate polymodels using a 5-year rolling window and monthly returns,[2] a set-up that has delivered results in the previous chapters. The use of 21 days of cumulative returns is motivated by the ambition of discovering slowly changing trends in the stock returns, which would be more difficult if we directly used daily data. The predictors are shifted by one month to eventually generate predictions for the next month. The original daily frequency of the data allows the predictions we provide to be recomputed every day, yet, the elementary models' parameters are

---

[1] Estimated with OLS.
[2] Cumulative returns of 21 days, with compounding.

only re-estimated at the beginning of each month, to save computation time keeping in mind the relatively deep window of 5 years that we use for the estimations.

Performing predictions of the specific stock returns using these techniques may seem like a brutal use of polymodels, however, it requires a very careful and detail-oriented design of each of the steps of the signal extraction. Therefore, we present in the first section of the chapter the methodological basis on which we construct our framework. We then propose a method for prediction selection, by evaluating different metrics for filtering, and providing a dynamic, point-in-time filter. In the third section, we propose a novel method for the aggregation of the predictions in a polymodel, which we benchmark with a more standard approach that is also presented. This is followed by a section dedicated to the assessment of the quality of the predictions and their associated methods for selection and aggregation. Such an evaluation is done using a trading strategy, which is complemented in the final section by a series of robustness tests.

6.2 Methodological Foundations

We describe below the methodological choices that have been made for the conception of the signal.

6.2.1 Data

The data used is the same as that used in Chaps. 4 and 5.

Recall that our investment universe is thus composed of the 500 largest market capitalizations in the US stock markets, restated monthly.

The data is split into two parts: some stock-level, cross-sectional data and a diversified set of time series of financial markets data.

As a reminder, stock-level data is composed of daily returns, market capitalizations, book-to-market ratios and industry classification as defined by level 3 of the GICS classification (Global Industry Classification Standard, MSCI), all measured using end-of-day data in USD. Our sample covers the period from 1999-01-01 to 2018-09-30.[3] For more details on the data used, the reader may refer to Chap. 5.

The other part of the data is composed of a diversified factor set of end-of-day returns including 1,134 variables. This factor set has a broad geographical coverage of all the developed and most of the emerging economies. Variables are defined at different geographical scales, some being worldwide, some regional, and some being country-related. Furthermore, all the most important asset classes are covered, including stock indexes, equity factor indices (i.e. classical risk premia), currencies,

---

[3] The stock-level data is provided by Refinitiv (ex-Reuters).

commodities, money market, sovereign bonds, corporate bonds, real estate, hedge funds and volatility indices. The dataset is defined between 1995-01-02 and 2018-10-16.[4] A complete description of the factor set is available in Chap. 4.

6.2.2   Time-series Predictions, Cross-sectional Portfolio Construction

Specific returns predictions are made by the elementary models in the time-series dimension. It is thus possible to assess the quality of the predictions for a particular stock. However, it is clear that the analysis is more robust when considered directly at an aggregated level, i.e. jointly for the 500 stock returns that are our target variables. Furthermore, as we validate the economic significance of the results by a trading strategy, we should also use a portfolio construction based on the aggregated predictions, as it would be the simplest way to assess the performance. Hence, we consider hereafter a long/short portfolio construction, in which the quality of the predictions is evaluated in a relative manner inside the cross-section of the stocks.

Conforming to the format that is used in the trading strategy, we always use a normalized version of the signal afterwards, which is ranked and set between -1 and $+1$. As mentioned in Chap. 5, this kind of normalization is quite common in the literature. The use of ranking leads us to mainly consider the ordinality of the predictions, since the normalized predictions are evenly spaced over their space of definition, while non-ranked signals incorporate an additional cardinality component, stating by how much one prediction is better than another. Considering the difficulty of the task of predicting the cross-section of stock returns, and for the purpose of robustness, we thus do not consider the cardinality dimension in the current chapter.

6.2.3   Subtracting the Average Return

Since polymodels are estimated over a rolling window, the average of the estimated model is equivalent to the average of the stock return inside the rolling window.

However, the average stock return over the past month is a strong predictor of the future returns. Jegadeesh (1990) observed a short-term reversal effect, in which the returns of the following month are negatively correlated with the returns of the past month. With Titman (1993), he also documented that this correlation becomes positive with the returns of the past year, an effect called momentum. Finally, the stock returns also show a form of long-term reversal (De Bondt & Thaler, 1985). So,

---

[4]The factor set data is mainly provided by Bloomberg.

there is a continuum of strong positive and negative correlations of the future returns with the past average returns computed over different time frames.

Hence, depending on the window used to estimate the elementary models, the average estimation inside the window reflects one of these effects, and thus positively or negatively affects the predictive power of the predictions made by the polymodel.

It is thus extremely important to subtract the average stock return from the predictions. This guarantees that the results of the predictions are not just a sophisticated re-invention of signals that are already well documented in the literature. Throughout the chapter, we thus only consider the predictions on top of this average return, and we refer to these corrected predictions simply as "predictions".

6.3 Predictions Selection

Here we compare different metrics to select the predictors. The purpose of this selection is of course to remove noise, since it is not a sustainable position to claim that all the 1,134 factors used for the predictions are linked with the 500 stock returns that we are trying to predict.

The selection may be done in several ways, but it should always be connected to some evaluation of the goodness of fit or of the statistical significance of the elementary models. Since there is no particular reason to select one of them *a priori*, we investigate several measures that may be used for prediction selection.

For a given polymodel, these measures are investigated by computing the average prediction of the current stock by all the selected factors in a particular quantile of the measure. For example, we may compute the average prediction for Microsoft returns of the top 60% of the factors in terms of the root mean squared error (RMSE). This computation is repeated among all the stocks being predicted, so that we build predictions of the cross-section of stock returns at each point in time. We then compute a full-sample panel regression of the stock returns of the next day[5] on the average predictions for a given quantile (60% in our example). The statistical significance of the predictions is simply assessed using the *t*-statistic of the parameter estimated. We finally repeat this procedure for several quantiles.

The regressions mentioned above between the normalized signal and the future returns should not be done on raw returns. As in the estimation of the polymodels, we want to measure the ability of the polymodels to predict the specific component of the raw returns. We thus use the future specific returns, still defined as the residuals of the cross-sectional regression.

We assume that at least a few predictors just bring noise to the average prediction. Our expectation is that if a measure used for selection is able to successfully distinguish the noise from the signal, then introducing more factors in the average

---

[5] The final trading strategy is expected to be rebalanced daily.

prediction (i.e., increasing the quantile of the selection) should first lead to an increase in t-statistics, up to a point after which the newly added factors carry so much noise that the significance decreases.

This methodology to evaluate prediction selection metrics is clear and simple, allowing for a first overall understanding of the question. Also, it makes different measures comparable, since the use of quantiles leads to the use of buckets of factors to compute the average predictions in a transparent manner.[6] This simplicity comes at the cost of making several assumptions. First, since we retain a selection metric based on full sample results to build a *rolling* trading strategy, we use future information. Such a practice can only be done by assuming that the quality of the different selection metrics is relatively constant over time, so that the current analysis may have been done with similar results at the beginning of our sample using a similar sample of past data. This assumption also implies that our choice of a selection metric also brings results out of sample. Secondly, we assume a weak form of homogeneity among the different polymodels. Indeed, since the quantile threshold used to compute the t-statistics is the same among polymodels, the robustness of our results depends on the fact that the polymodels behave approximately in the same manner at different quantiles. Although this is a simplification of reality, these assumptions seem acceptable as they also simplify the problem that we are facing.

6.3.1 Root Mean Squared Error Filter Evaluation

The root mean squared error (RMSE) may be the most basic measure of the goodness of fit of a model. At a given point in time and for a given predictor, it is defined as:

$$RMSE_{a,i,t} = \sqrt{\frac{1}{\tau} \sum_{s=t-\tau}^{t} \left(y_{a,s} - LNLM_{a,i}(x_{i,s}) \right)^2}. \tag{6.3}$$

Here "τ" is the number of observations available in the rolling window at date t, "i" is the index of the predictor, and "a" is the index of the stock (if not specified, all the equations below are presented for a given polymodel).

The RMSE is computed inside of its particular rolling window for each elementary model of each of the polymodels. Below are the t-statistics that reflect the predictive power of the average predictions for different quantiles of the RMSE (the smallest quantile being the highest level of goodness of fit) (Fig. 6.1).

The filter clearly shows the expected pattern for a proper filter, since we can see that there is a decrease in the predictive power of the aggregated signal when the worst factors in terms of RMSE are used in the prediction.

---

[6]i.e. with the same number of factors in each bucket, whatever the metric.

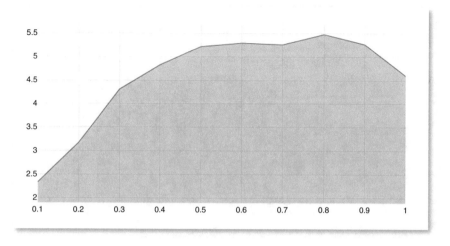

Fig. 6.1   RMSE Filter: t-statistic as a function of the noise filter quantile

6.3.2   p-*value Filter Evaluation*

The p-value is perhaps the most common measure of statistical significance. It may be computed from an F-test comparing the lack-of-fit sum of squares to the pure error sum of squares. The ratio of these two quantities, weighted by their degrees of freedom, is the F-statistic that is replaced in the cumulative distribution function of a Fisher–Snedecor distribution to get the p-value:

$$F = \frac{lack - of - fit\ sum\ of squares/degrees\ of\ freedom}{pure - error\ sum\ of\ squares/degrees\ of\ freedom}. \tag{6.4}$$

Thus, the p-value computed from an F-test is fundamentally different from the RMSE, since in addition to considering the goodness-of-fit that is achieved, it also considers the goodness-of-fit that is achievable.

Below are the t-statistics attained by the full sample regression for different quantiles of filtering (Fig. 6.2):

Considering this graph, the p-value computed from an F-test does not seem to be usable as a reliable noise filter. Indeed, even the worst predictions in terms of p-value still add value to the aggregated prediction. The mismatch of the p-value in the context of filtering may be due to the violation of the assumption that the errors of the estimated elementary models are independently and normally distributed. One should also consider that the p-value may be a more suitable metric to weight the different predictions rather than to select them.

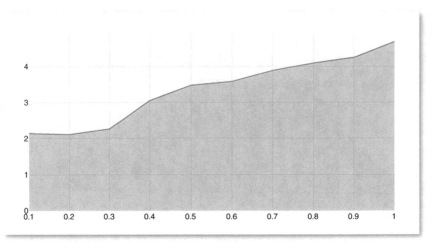

Fig. 6.2 *p*-value Filter: *t*-statistic as a function of the noise filter quantile

6.3.3   Bayesian Information Criterion Filter Evaluation

The Bayesian Information Criterion (BIC) is defined as:

$$BIC = \ln{(q)}\rho - 2\ln{\left(\widehat{L}\right)}. \tag{6.5}$$

Here "q" is the number of observations, " ρ " is the number of parameters estimated by the model, and "$\widehat{L}$" is the maximized value of the likelihood function.

Since the estimation of the likelihood function is a non-trivial task, and considering the computational intensity of the polymodels estimation, we assume the errors of the model to be independently, identically and normally distributed. This Gaussian assumption leads to the following modification of the BIC formula:

$$BIC = \ln{(q)}\rho + q \times \ln{\left(\frac{1}{q}\sum_{d=1}^{q}\left(y_{a,d} - LNLM_{a,i}(x_{i,d})\right)^{2}\right)}. \tag{6.6}$$

Here "d" is the observation index.

We can see that under this assumption the definition of BIC becomes close to that of RMSE. Still, the first term of Eq. (6.6) accounts for and penalizes the number of observations available for the fit. Recall that the polymodel estimation is done using elementary models that may have a different number of observations, depending on the sample size of each regressor in the current window. Taking into account this particular feature of the polymodels is thus an advantage of BIC over RMSE to discriminate among different elementary models.

Not surprisingly, the results reached by the BIC filter are comparable to those obtained using RMSE filtering (Fig. 6.3):

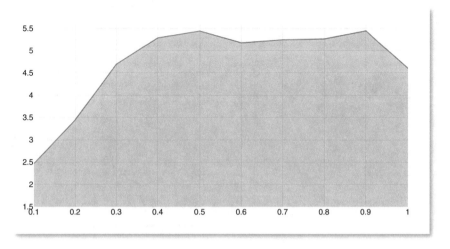

Fig. 6.3 BIC filter: t-statistic as a function of the noise filter quantile

6.3.4   Selection Using Dynamic Optimal Filtering

Based on the full sample results of our assessments of various metrics for filtering, it is clear that the F-test-based p-value should not be used to perform a selection of the predictions, while RMSE and BIC have acceptable performances. Since these performances, along with the mathematical definitions of these metrics, are comparable, we choose to avoid any arbitrary choice by combining them. Because we ultimately rely on quantiles, we can create a balanced combined metric by summing the ranked values of BIC and RMSE and ranking this sum:

$$DFM_t = rank[rank(BIC_t) + rank(RMSE_t)]. \qquad (6.7)$$

Here DFM is the vector of "double filtering metric" for all elementary models at a time t, and rank is the ranking operator.

We need to determine an appropriate quantile threshold for filtering. However, although the use of full sample data is considered here as a tolerable way to choose among metrics for filtering,[7] keeping a full sample approach to set a threshold would induce a high risk of overfitting, since the final results would reflect some form of in-sample optimization. The setting of the threshold for filtering is thus only done using point-in-time data. Also, the threshold is adapted to each of the polymodels, so that there is no longer a homogeneity assumption among the different stock returns.

For a particular polymodel, at each date we compute the average predictions filtered by different quantiles of the double filtering metric. Then, each quarter, we re-compute the univariate regression of the next day stock returns on the current

---

[7]Their effectiveness being considered as not time-dependent.

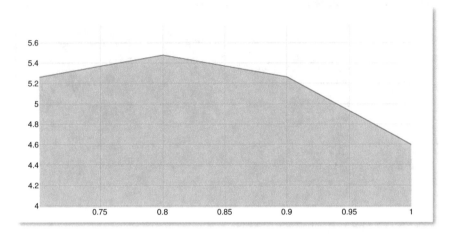

Fig. 6.4 Example of DFM filter: t-statistic as a function of the noise filter quantile

filtered/aggregated prediction, for different quantiles. For the current date, we thus obtain again a t-statistics curve that reflects the predictive power of different filtering quantiles, which may look like the Fig. 6.4, presented as a representative example of the problem:

For the current polymodel and the current date, we obtain an assessment of the predictive power of the different thresholds possible. We are thereby able to choose a threshold that is optimal in some sense relative to that curve. Note that since it does not seem that the selected metrics can do more than marginally remove the noise, we restrict the analysis to the highest quantiles,[8] which helps to avoid being trapped in some drastically scarce (and spurious) optima.

A simple and maybe naive approach would be to simply select the quantile that corresponds to the highest t-statistic. We proceed differently.

When new predictors are included by increasing the quantile, we may expect the following outputs:

- If the added information is not included in the previous quantiles (so the added predictors are uncorrelated) and is relevant, the average prediction should show a higher t-statistic.
- If the added information is not included in the previous quantiles, and is only *weakly* relevant, the overall, *unweighted* aggregated predictive power may not be as good, leading to a smaller t-statistic.
- If this information is relevant but already partly included in the previous quantiles, it may then become over-weighted in the aggregate prediction, leading to a smaller t-statistic.

---

[8] From 70% to 100%.

- Finally, if this information is pure noise coming from spurious predictors, we can expect the *t*-statistic to drop sharply.

From these propositions describing the effect of increasing the quantile on the *t*-statistic, we may consider that a decrease in the *t*-statistic does not necessarily imply that the added predictors contain only noise. Since information is precious, we follow a conservative approach, only removing predictions that we most suspect to be pure noise. Eventually, it seems more likely that the largest addition of noise after a quantile increase would occur after the largest drop in *t*-statistic. Hence, we define the optimal threshold as the threshold that minimizes the first derivative of the *t*-statistic curve.

The optimal threshold is then re-computed dynamically, each quarter, using regressions performed with a 36-month rolling window. This procedure is applied separately to each of the polymodels.

6.4   Aggregation of Predictions

Using the predictions selected by the dynamic optimal filter, we present in this section several candidate techniques to aggregate the predictions.

6.4.1   Prediction Aggregation Using Bayesian Model Averaging

Bayesian Model Averaging (BMA) is a commonly used method to combine forecasts (for an introduction, see Hoeting et al., 1999 and Raftery et al., 1997). It has been used in the context of polymodels for the combination of elementary models by Guan (2019). By associating a probability to each elementary model, BMA proposes a combined prediction that takes into account model uncertainty. For a given polymodel:

$$E\left[y_{t+1}|X_t\right] = \sum_{i=1}^{n} \widehat{\varphi_{i,t}}(X_{i,t}) \times p\left(\varphi_{i,t}(X_{i,t})|X_t\right) \tag{6.8}$$

Here y_{t+1} is the variable we are trying to predict, X_t is the factor set at time t, and $\varphi_i(X_{i,t})$ is the elementary model of factor i at time t.

Such an approach is quite intuitive, as the different predictions are believability weighted, however, it does not take into account that the different models may be correlated. The (posterior) probabilities are defined by Bayes' theorem as:

$$p\big(\varphi_{i,t}(X_{i,t})|X_t\big) = \frac{p\big(X_t|\varphi_{i,t}(X_{i,t})\big) * p\big(\varphi_{i,t}(X_{i,t})\big)}{\sum_{j=1}^{I} p\big(X_t|\varphi_{j,t}(X_{j,t})\big) * p\big(\varphi_{j,t}(X_{j,t})\big)}. \qquad (6.9)$$

As the posterior probability is expressed in terms of likelihood, it is related to BIC, which can be used to approximate it (Raftery, 1995):

$$p\big(\varphi_{i,t}(X_{i,t})|X_t\big) \approx \frac{e^{-BIC_i/2}}{\sum_{j=1}^{n} e^{-BIC_j/2}}. \qquad (6.10)$$

This approximation is required in our case because of the computational intensity of the polymodel estimation (recall that BIC is very easy to compute as a function of the squared errors, under some assumptions).

Since BIC can be very large, in practice, we can subtract from it any convenient value, usually the minimum of the BIC values of the different models, which does not impact the ratio. In our case, the resulting dispersion of the BIC values is still very large, leading to the (numerical) failure of computing reliable exponentiations. We thus approximate the above ratio further by using:

$$p\big(\varphi_{i,t}(X_{i,t})|X_t\big) \approx \frac{-BIC_i}{\sum_{j=1}^{n} -BIC_j}. \qquad (6.11)$$

This final approximation leads to a concentration of the probabilities, which may weaken the results of BMA.

6.4.2   Prediction Aggregation Using Added Value Averaging

We introduce here a weighting method that aims at averaging the predictions as a function of their added value. To explain what we mean by this term, let us re-use the amphitheater metaphor of Chap. 2. We are in the amphitheater of a university, in front of selected students (filtered predictors), who try to predict the future returns of an asset over several rounds. We observe their repeated guesses, and we need to aggregate them to obtain a combined prediction. We propose that the weights of the individual predictions should consider the two following properties of the said predictions:

- Believability: students who tend to provide better predictions over time should have a larger weight.
- Originality: students who provide different predictions than others should also have a larger weight, since someone who just repeats what is already well known does not add any information.

Note that these two properties are not additive, but multiplicative. Someone who is believable but not original should have a small weight, as their opinion is repeated (if you get exactly the same prediction 10 times, you would like to weight each of them at something like 1/10 before summing them). Someone who is original but not believable is not someone that you want to consider and thus should have a very small weight. When both of these properties are simultaneously present for a predictor, then we consider the predictor to *add value* to the aggregate prediction, and hence they should be associated with a large weight.

When thinking about believability in terms of the usual tools we use, we would like to select a classical goodness of fit metric for the elementary models. But in terms of the metaphor of the amphitheater, we would like to consider the accuracy of each student's prediction. Using Bayesian terms, the first perspective on the problem is *a priori*, and the second one is *a posteriori*. The first perspective considers in-sample goodness of fit, and the second out-of-sample results. In our usual concern to fight against overfitting, we favor the second approach, and propose to use an out-of-sample version of RMSE:

$$RMSE_{OOS,i,t} = \sqrt{\frac{1}{\tau}\sum_{s=t-\tau}^{t}\left(y_s - \varphi_{i,s-21}(x_{i,s-21})\right)^2}. \qquad (6.12)$$

Here "i" is the index of the elementary model considered, "t" is a given date, " τ " is the length of the window used for computation, "y_t" is the realized stock return, and " $LNLM_{i,\,t\,-\,21}(x_{i,\,t\,-\,21})$" is the prediction of this stock return made a month ago.

In addition, to consider the effective predictive power of our factors, this measure has the advantage that it can be computed on shorter time frames than the window of the polymodels. Indeed, providing robust estimates of the elementary models often requires deep rolling windows for the polymodel estimation, usually set to 5 years (i.e. roughly 1,260 points). Such a depth makes sense if we consider that we always use a 10-fold cross-validation in the estimation of the LNLM model, which is thus backed by pseudo-out-of-sample selections of μ^9 performed on 126 points. This 3-digit number may be a *minimum* to guarantee the reliability of the estimations. Nevertheless, if we need such a window depth to provide a good model of the link between the target variables and the predictors, there is no reason to consider *a priori* that this window is the most suitable to assess the significance of such a link. The dynamics of the links between predictors and target variables may be changing more quickly, and in the current application, we therefore consider a 2-year window for the computation of the out-of-sample RMSE.

The smaller the RMSE, the larger the believability of the predictor. Our believability measure is thus defined as:

---

[9]The non-linearity propensity parameter of the LNLM model.

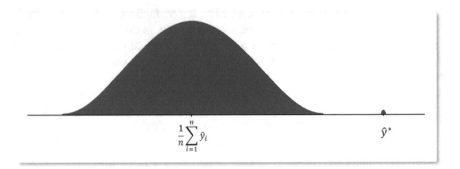

Fig. 6.5 Stylized representation of the distribution of predictions

$$B_{i,t} \overset{\text{set}}{=} log \left(RMSE_{OOS,i,t}\right)^2. \tag{6.13}$$

Shannon (1948) introduced a measure of information of a message m named "self-information" or "surprisal", defined as a function of its probability of occurence $p()$ such as:

$$Info(m) = - log \left(p(m)\right). \tag{6.14}$$

This definition may not be straightforward to understand. Returning to the amphitheater metaphor, let us consider the distribution of the predictions $\widehat{y}_i$ we get at a certain time t (Fig. 6.5):

Considering all the predictions to be relevant (they have been filtered), a prediction equal to the mean $\frac{1}{n}\sum_{i=1}^{n}\widehat{y}_i$ of the distribution does not contain much information because, while taking into account all the other predictions, we already know that the m value is possible, and even probable. On the other hand, the point "$\widehat{y}^{*}$" contains a lot of information, because if we don't have this point, we don't know that the event "$\widehat{y}^{*}$" is even possible. Thus, a relevant message about an event that is already known for sure ($p = 1$) does not carry any information, while a relevant message about an event that is totally unexpected carries a lot of information. Hence, in the amphitheater, a student, even good, that just repeats what others have already said, does not bring much information, while a good student that predicts something totally unexpected should be listen to carefully.

In our context, Shannon's measure of information reflects the originality of the prediction compared to the distribution of the predictions among elementary models. We thus propose the following measure of originality of a prediction:

$$o\left(\widehat{\varphi}_{i,t}(x_{i,t})\right) \overset{\text{set}}{=} log \left(p\left(\widehat{\varphi}_{i,t}(x_{i,t}) \mid \{\widehat{\varphi}_{j,t}(x_{j,t}) \forall j \in [1:n]\} \right) \right)^2. \tag{6.15}$$

We replaced the use of the negation by a square to get a positive value of the logarithm because it allows more extremized measures of originality, leading to a more differentiating metric.

To consider a student to be original, it is not sufficient to observe her prediction at a single point in time, the predictions should be repeatedly original. Therefore our final originality measure of a predictor is:

$$O_{i,t} \stackrel{\text{def}}{=} \frac{1}{\tau} \sum_{s=t-\tau}^{t} o\left(\widehat{\varphi}_{i,s}(x_{i,s})\right). \tag{6.16}$$

We assume that the dynamics of the originality property is tied to the dynamics of the elementary models' estimations, hence we use a 5-year rolling window to compute the originality measures of predictors.

As stated previously, a predictor adds value to the overall prediction if it is simultaneously believable and original, so our added value measure is defined as:

$$AV_{i,t} \stackrel{\text{def}}{=} B_{i,t} \times O_{i,t}. \tag{6.17}$$

The aggregate predictions made using Added Value Averaging (AVA) are:

$$E\left[y_{t+1}|X_t\right] = \sum_{i=1}^{n} \widehat{\varphi}_{i,t}(X_{i,t}) \times AV_{i,t}. \tag{6.18}$$

6.4.3   Uncertainty of Aggregate Predictions

This section does not directly concern the way we aggregate predictions, but how we weight the cross-section of the aggregated predictions. We propose the following: when there is high dispersion in the polymodels' predictions of a given stock return, it reflects an uncertainty about the aggregated prediction, which should be penalized.

We thus recommend that each of the aggregated predictions should be divided by the standard deviation of the polymodels' predictions at each date.

Note that this would not be an improvement of the predictions themselves, since it relates to the confidence we have in each aggregated prediction, it would be more an artefact of the cross-sectional trading strategy that we use to value our predictions.

6.5   Trading Strategy

6.5.1   Methodology

As for the predictions' estimations, the investment universe used for the trading strategy is roughly equivalent to the S&P 500, as we select the 500 largest market capitalizations that are traded in the US stock markets, restated monthly.

As in the previous chapter, we consider a long/short trading strategy. Recall that we construct a score that is based on the cross-section of the aggregated predictions of the polymodel of each stock. The score is sorted in ascending order, ranked and normalized to be uniformly distributed between -1 and $+1$. As stated previously, this allows us to only consider the ability of the strategy to predict the stock returns on a relative basis.

We still use the characteristic portfolio of the score to obtain the positions[10] of the portfolio. As a reminder, the characteristic portfolio is the minimum-variance portfolio that has a unit exposure to the score (Grinold & Kahn, 2000). The strategy performance is then simply the sum of the daily returns of each of the individual positions in our investment universe.

There is no control for the leverage of the portfolio positions, but all the different variations of the strategy's returns are normalized in order to exhibit a full sample yearly volatility of 10%, which allows for their comparison (the choice of a full sample volatility normalization may be less realistic than a rolling normalization, however it guarantees that local spikes of volatility, e.g. during drawdowns of the strategy, are not hidden by volatility smoothing).

Below, we first propose a version of the strategy that combines all the steps of the prediction's computation (filtering, aggregation, controlling for stocks' uncertainty). We also evaluate the statistical significance of the Sharpe ratio by computing the p-value of this metric.

We then disentangle the different techniques used to compute the aggregate predictions and assess their cumulative additivity to the strategy's performance.

6.5.2   Performance

The final version we selected is filtered by the Dynamic Optimal Filter presented above, weighted using the Added Value Averaging technique, and taking into account each aggregate prediction uncertainty.

The graph below presents the cumulative performance (compounded) of the strategy, displayed in log-scale (Fig. 6.6):

The strategy achieves a Sharpe ratio of 0.91, which is economically significant.

---

[10] As in the previous chapter, the positions are lagged by two days for the computation of the portfolio returns.

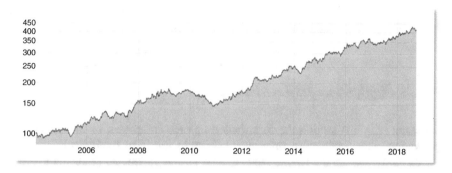

Fig. 6.6   Polymodel's predictions strategy performance (10% annual volatility)

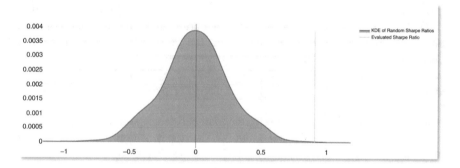

Fig. 6.7   Distribution of random Sharpe ratios

The statistical significance of this metric is evaluated by computing its p-value. To do this, we follow the procedure described below:

- We first compute the Sharpe ratio achieved by the strategy.
- Secondly, we shuffle the individual stock returns, and recompute the portfolio performance, from which we deduce a Sharpe ratio. Since the returns are shuffled, this new value only reflects randomness, although it preserves the autocorrelation structure of the signal.
- The previous step is repeated 300 times, so that we get a distribution of random Sharpe ratios.
- This distribution is smoothed using a kernel density estimate (Parzen, 1962; Rosenblatt, 1956), which allows us to reduce threshold effects, increase the accuracy, and slightly extend the interval on which the distribution is defined.
- The Sharpe ratio of the initial portfolio is then replaced in the distribution, the p-value of the Sharpe ratio being equal to the integral of the probabilities of the random Sharpe ratios beyond that point.

Below is the distribution obtained with this methodology (Fig. 6.7):

The *p*-value of the Sharpe ratio of the final strategy is equal to 0.02%, which is highly statistically significant.

6.5.3   Significance of the Methods

We explore in this section the contribution of each component of the aggregated prediction to the final performance and evaluate its statistical significance.

We first compare the simple mean among all the polymodel predictions to the simple mean filtered using the dynamic optimal filter (Fig. 6.8):

The filtering allows the Sharpe ratio to increase from 0.5 to 0.63. While regressing the returns of the filtered portfolio on the unfiltered portfolio, we note that the constant of the model is positive, though weakly statistically significant (Table 6.1):

We may expect the filter used to produce a more (statistically) significant increase in the performance. However, if we examine the performances above, it seems to induce a reduction of the large drawdowns of the strategy. The maximum drawdown of the raw strategy indeed decreases from -35% to -28%, and it is clear from the graph of the underwater drawdowns that the filter provides a kind of downside protection (Fig. 6.9):

Based on these findings, the use of the filter still seems justified.

Secondly, we compare the BMA and AVA methods, with the simple mean as a benchmark, using filtered predictions in all the cases (Fig. 6.10):

The Sharpe ratios achieved are respectively 0.75, 0.63 and 0.61 for the AVA, simple mean and BMA methods. Additionally, it is clear while comparing regression

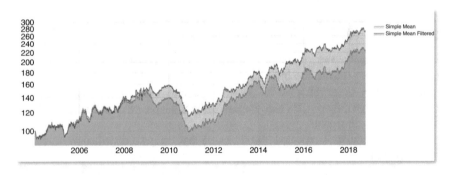

Fig. 6.8 Comparison of performances for simple mean and simple mean Filtered

Table 6.1 Regression of the simple mean filtered returns on the simple mean returns

	Coef	Std err	t	$P > |t|$	[95.0% Conf. Int.]	
Constant	6.78E−05	0.0000	1.883	6.00% (*)	−2.81e−06	0.000
Simple mean	0.9376	0.006	164.224	0.00% (***)	0.926	0.949

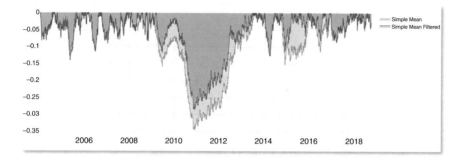

Fig. 6.9 Underwater drawdowns of simple mean and simple mean filtered

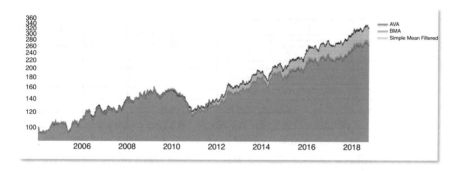

Fig. 6.10 Comparison of performances for AVA, BMA, and simple mean filtered

Table 6.2 Regressions of AVA and BMA returns on the simple mean filtered returns

	Variables	Coef	Std err	t	P > |t|	[95.0% Conf. Int.]
AVA	Const	5.30E−05	0.0000	2.548	1.10% (**)	1.22e−05 9.37e−05
	Simple mean filtered	9.80E−01	0.0030	297.134	0.00% (***)	0.973 0.986
BMA	Const	−5.99E−06	0.0000	−0.569	57.00% ( )	−2.66e−05 1.47e−05
	Simple mean filtered	9.95E−01	0.0020	595.39	0.00% (***)	0.992 0.998

tables (Table 6.2) that the AVA provides a statistically significant extra-return on top of the simple averaging, which is not the case for BMA:

There are several possible explanations for the failure of BMA in our context. First, it has been developed to merge models that include a different number of parameters, which is not the case for our elementary models, for which the BIC only penalizes the number of observations, this being comparable for most of the models.

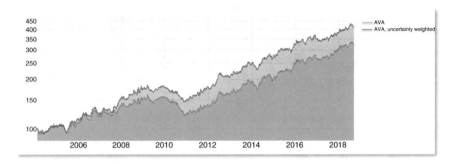

Fig. 6.11 Comparison of performances for AVA and AVA uncertainty weighted

Table 6.3 Regression of the AVA uncertainty weighted returns on the AVA returns

	Coef	Std err	t	$P > |t|$	[95.0% Conf. Int.]	
Const	8.35E−05	0.0000	2.454	1.40% (**)	1.68e−05	0.000
AVA	9.45E−01	0.0050	175.133	0.00% (***)	0.934	0.955

BMA is thus not as good at discriminating among elementary models, since it just focusses on the goodness of fit, while AVA also takes into account the correlations among the different predictions. The BIC approximation of the posterior, although unavoidable in our context, may lead to a weak estimation of the likelihood, as it requires assumptions. Furthermore, the BIC is obtained using assumptions on the distribution of the errors. Finally, we further approximated Raftery's BIC approximation by using the non-exponentiated ratio of the BICs. Hence, it seems that the failure of the BMA in the context of extensive polymodel computations is more due to practical infeasibilities than to a problem with the method itself, and it would certainly not be possible to claim the opposite.

Finally, taking into account the uncertainty of the aggregate predictions in the cross-sectional strategy is also adding value, the Sharpe ratio increasing from 0.75 to 0.91 (Fig. 6.11):

In addition to being economically significant, this last technique is also statistically significant (Table 6.3):

These results indicate that the performance of our strategy is significant, and that it is due to the combination of different techniques all of which already have a certain predictive power. We thus control further for the reliability of these results by performing a series of robustness tests.

6.6   Robustness Tests

6.6.1   *Correlations with Standard Factors*

Since we have adopted a cross-sectional portfolio construction, we have to control for the possibility that we have simply proposed a sophisticated re-invention of risk premia or factors already well documented in the literature. For this purpose, we control for the Fama–French 3 factors (1993), being the market, the value and size,[11] the standard momentum, as defined by Carhart (1997), the (short-term) reversal observed by Jegadeesh (1990), the long-term reversal of De Bondt and Thaler (1985) and the low volatility anomaly (Baker et al., 2011).

The portfolio positions of these different anomalies are computed using the same long/short portfolio construction as the polymodels' specific predictions, except for the market, for which the portfolio positions are simply equal to, for a given stock "a":

$$w_a = \frac{MarketCapitalization_a}{\sum MarketCapitalization} \qquad (6.19)$$

First, we display the correlation matrix of the portfolio returns of the anomalies listed above (Fig. 6.12):

The correlations among the returns of the polymodels' predictions and the returns of the other factors being extremely low, it is unlikely that the performance of our long/short strategy can be explained by the performance of other anomalies. Still, we control for it using a full sample regression (Table 6.4):

The regression shows that there is a statistically significant alpha which is left unexplained by the classical factors. The value of the constant corresponds roughly to a 10% abnormal return annually (as before, all the returns have been standardized to 10% of annual volatility). Although significant, the coefficients of the market and the low-volatility anomaly are of small magnitude. Overall, the R^2 of the regression is 1.2%: the standard factors do not explain the main part of the polymodels' predictions returns.

6.6.2   *Sensitivity to the Filter Regression Window*

The dynamic optimal filter we proposed uses a *t*-statistics curve to evaluate its objective function. These *t*-statistics have been obtained using some univariate regression of the stock returns on the filtered aggregated predictions (with different quantiles of filtering). These regressions are performed dynamically, using a rolling

---

[11] Defined here respectively as: a market-capitalization weighted average of returns; characteristics defined as the book to market ratio; and minus the logarithm of the market capitalization.

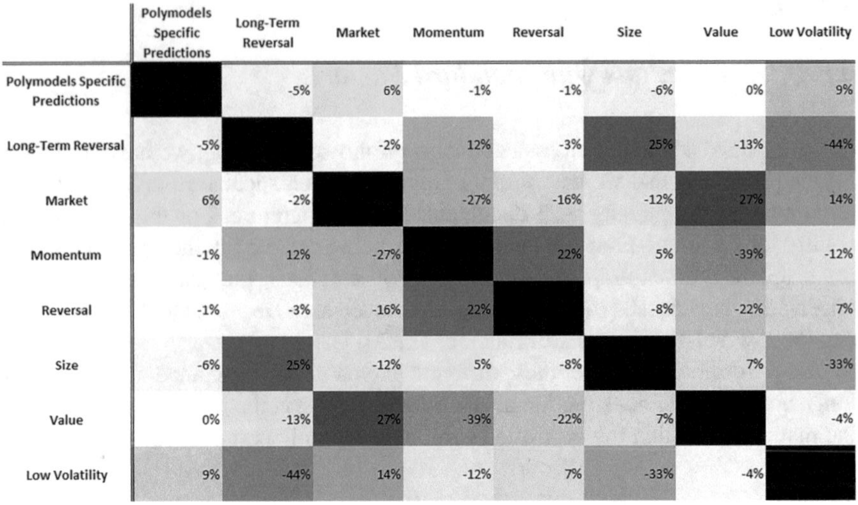

	Polymodels Specific Predictions	Long-Term Reversal	Market	Momentum	Reversal	Size	Value	Low Volatility
Polymodels Specific Predictions		-5%	6%	-1%	-1%	-6%	0%	9%
Long-Term Reversal	-5%		-2%	12%	-3%	25%	-13%	-44%
Market	6%	-2%		-27%	-16%	-12%	27%	14%
Momentum	-1%	12%	-27%		22%	5%	-39%	-12%
Reversal	-1%	-3%	-16%	22%		-8%	-22%	7%
Size	-6%	25%	-12%	5%	-8%		7%	-33%
Value	0%	-13%	27%	-39%	-22%	7%		-4%
Low Volatility	9%	-44%	14%	-12%	7%	-33%	-4%	

Fig. 6.12 Correlation matrix of the portfolio returns of several anomalies

Table 6.4 Regression of the polymodel's predictions returns on other factor returns

	Coef	Std err	t	$P > |t|$	[95.0% Conf. Int.]	
Const	0.0004	0	3.402	0.10% (***)	0.000	0.001
Long-term reversal	−0.0168	0.019	−0.894	37.10% ( )	−0.054	0.020
Market	0.0483	0.018	2.728	0.60% (***)	0.014	0.083
Momentum	0.0166	0.018	0.9	36.80% ( )	−0.020	0.053
Reversal	−0.0207	0.017	−1.205	22.80% ( )	−0.054	0.013
Size	−0.0281	0.018	−1.596	11.10% ( )	−0.063	0.006
Value	−0.0078	0.019	−0.421	67.40% ( )	−0.044	0.029
Low volatility	0.0719	0.019	3.729	0.00% (***)	0.034	0.110

window of 36 months. Below we control for the sensitivity of the strategy performance[12] to the choice of this particular window (Table 6.5):

The particular choice of a 36-month window does not really seem to impact the performance. Actually, the performance of the strategy may even be improved by a larger exploration of the parameter space, as the performance of the 48-month window suggests.

---

[12] Some small discrepancies may be observed among the reference figures in the different tables of sensibilities (e.g. here the Sharpe ratio for 36 months is 0.90 instead of the previous 0.91). This is due to the fact that we only use the returns for which all the dates are available for the different variations of the strategy. Although this approach is the fairest, because we don't want to compare Sharpe ratios computed over different periods, it produces discrepancies among tables, because the smallest common set of dates is different for each table, since it is determined by the particular depth of the windows we are testing.

Table 6.5 Sensitivity of the performance to the filter regression window

Dynamic optimal filter window	Average gross return (%)	Sharpe ratio	Turnover	Volatility (%)	Worst drawdown (%)
24 months	10.23	0.90	250.86	10.00	−21.01
30 months	10.31	0.91	250.86	10.00	−21.89
36 months	10.19	0.90	250.86	10.00	−21.46
42 months	10.27	0.91	250.86	10.00	−21.96
48 months	10.56	0.94	250.86	10.00	−21.80

6.6.3   Stability of the Performance

We analyze the stability of the performance through different delays to achieve the desired positions. The delay used in the rest of the chapter is of one day after the computation of the signal, which is conservative, since it assumes that an entire trading day is needed to reach the positions. This assumption may be realistic in a variety of cases, i.e. if the execution system used only sends orders at a particular time of the day (for example during the closing auction), or if orders are executed during the entire trading day in order to minimize the market impact. The delays used to assess the stability of the performance are of 1, 2, 3, 5, 10 and 21 days, and we approach it across several metrics (Table 6.6):

The Sharpe ratio consequently decreases after only a few days of delay to reach the positions. This usually indicates a fast signal, which is confirmed by the high level of the turnover,[13] as well as the autocorrelogram of the signal (Fig. 6.13):

Such a pattern implies that the strategy would be associated with substantial transactions costs, which should encourage us to combine it with other strategies in the case of an implementation.

Table 6.6 Performance statistics for different trading delays

Delay	Average gross return (%)	Sharpe ratio	Turnover	Volatility (%)	Worst drawdown (%)
1 Day	10.19	0.91	250.83	10.00	−21.46
2 Days	9.36	0.82	250.80	10.00	−22.32
3 Days	7.73	0.65	250.75	10.00	−26.49
5 Days	9.22	0.80	250.69	10.00	−19.48
10 Days	3.13	0.19	250.54	10.00	−27.40
21 Days	3.43	0.22	250.23	10.00	−26.04

---

[13] The turnover is computed as the number of rotations of the capital of the strategy per year.

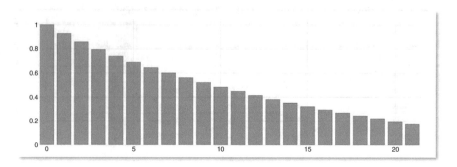

Fig. 6.13 Autocorrelogram of the signal

6.6.4   Sensitivity to the Believability Measure Window

In the computation of the weights used by the Added Value Averaging method, we compute the out-of-sample RMSE to derive a believability measure. The out-of-sample RMSE is obtained by comparing each of the predictions made by the predictors to the value of the stock returns that is finally achieved. The RMSE is obtained from the sum of the associated errors, and this sum is computed over a short-term rolling window of 24 months. Below, we test the impact on performance of different values of this parameter (Table 6.7):

Here, if the window seems relatively well chosen, it is clear that the parameter choice only has a weak impact on the strategy's performance.

6.6.5   Sensitivity to the Originality Measure Window

The other component of the Added Value Averaging predictions' weights is the originality measure, for which we compute the average information using a rolling 5-year window. The length of this rolling window has a negligible impact on the performance (Table 6.8):

Table 6.7 Sensitivity of the performance to the believability measure window

AVA—Believability window	Average gross return (%)	Sharpe ratio	Turnover	Volatility (%)	Worst drawdown (%)
12 months	10.08	0.89	250.86	10.00	−24.38
18 months	10.14	0.89	250.86	10.00	−22.47
24 months	10.19	0.90	250.86	10.00	−21.46
30 months	10.16	0.90	250.86	10.00	−21.94
36 months	10.02	0.88	250.86	10.00	−22.02

Table 6.8 Sensitivity of the performance to the originality measure window

AVA—Originality window	Average gross return (%)	Sharpe ratio	Turnover	Volatility (%)	Worst drawdown (%)
3 years	10.20	0.90	250.86	10.00	−21.49
4 years	10.20	0.90	250.86	10.00	−21.30
5 years	10.19	0.90	250.86	10.00	−21.46
6 years	10.16	0.90	250.86	10.00	−21.64
7 years	10.10	0.89	250.86	10.00	−21.85

6.6.6 Unaccounted Parameter Sensitivities

Two parameters may be tested for their sensitivities and are left unaccounted in the current series of robustness tests, because of the impossibility of re-performing the polymodel estimations due to the extensive computations they require.

The window used by the polymodels estimation has been set to a default of 5 years. Such a choice may have an impact, as we saw in Chap. 4. However, both Chaps. 4 and 5 seem to indicate that a 5-year window to estimate the polymodel is the best option. As explained previously, this particular window uses 1,260 points, so 126 are used in each of the pseudo-out-of-sample estimations of LNLM's μ by stratified 10-fold cross-validation. The 5-year window may lead to the optimal balance between a robust parameter choice and a sufficiently adaptive modeling of the asset returns dynamics.

Recall that in the stratified 10-fold cross-validation of LNLM, the observations are drawn randomly from each observations' distribution layer. The impact of this random drawing should be controlled by a sensitivity analysis to the seed used. If it is also not possible to test for the seed sensitivity because of computational resources, the results of Chap. 5 show that the LNLM fit does not seem very sensitive to the seed.

6.7 Conclusions

In this chapter, we proposed reliable predictions of the specific component of the stock returns. These predictions were validated by a successful trading strategy, which reached a Sharpe ratio of 0.91 (p-value 0.02%). We also demonstrated that the signal is distinct from the standard factors used to explain the cross-section of the stock returns.

These results have been achieved thanks to the development of new techniques that answer the common concerns about the use of polymodel predictions: the selection and the aggregation of the predictions.

For the purpose of selection, we proposed a dynamic and point-in-time filter that successfully removes the noisiest predictors and reduces the drawdowns of the trading strategy.

We proposed a novel method for the aggregation that simultaneously takes into account the pseudo-out-of-sample goodness of fit of the predictors and the correlations among the predictions. The Added Value Averaging method has been shown to significantly increase the performance of the trading strategy.

Along with the market returns forecasts of Chap. 4 and the industry returns forecasts of Chap. 5, the forecasts of the current chapter form the third base pillar of a multi-dimensional prediction system for the equity markets. These results further support the reliability of polymodels as a tool to analyze nodes of complex systems, make predictions, and ultimately design profitable trading strategies.

References

Baker, M., Bradley, B., & Wurgler, J. (2011). Benchmarks as limits to arbitrage: Understanding the low-volatility anomaly. *Financial Analysts Journal, 67(1)*, 40–54.

Carhart, M. M. (1997). On persistence in mutual fund performance. *The Journal of finance, 52(1)*, 57–82.

De Bondt, W. F., & Thaler, R. (1985). Does the stock market overreact? *The Journal of finance, 40(3)*, 793–805.

Fama, E. F., & French, K. R. (1993). Common risk factors in the returns on stocks and bonds. *Journal of Financial Economics, 33(1)*, 3–56.

Grinold, R. C., & Kahn, R. N. (2000). *Active portfolio management: A quantitative approach for producing superior returns and controlling risk.* McGraw-Hill.

Guan, Y. (2019). *Polymodel: Application in risk assessment and portfolio construction* (Doctoral dissertation). State University of New York at Stony Brook.

Hoeting, J. A., Madigan, D., Raftery, A. E., & Volinsky, C. T. (1999). Bayesian model averaging: A tutorial. *Statistical Science, 14(4)*, 382–401.

Jegadeesh, N. (1990). Evidence of predictable behavior of security returns. *The Journal of Finance, 45(3)*, 881–898.

Jegadeesh, N., & Titman, S. (1993). Returns to buying winners and selling losers: Implications for stock market efficiency. *The Journal of Finance, 48(1)*, 65–91.

King, B. F. (1966). Market and industry factors in stock price behavior. *The Journal of Business, 39(1)*, 139–190.

Parzen, E. (1962). On estimation of a probability density function and mode. *The Annals of Mathematical Statistics, 33(3)*, 1065–1076.

Raftery, A. E. (1995). Bayesian model selection in social research. *Sociological Methodology, 25*, 111–163.

Raftery, A. E., Madigan, D., & Hoeting, J. A. (1997). Bayesian model averaging for linear regression models. *Journal of the American Statistical Association, 92(437)*, 179–191.

Roll, R. (1988). R2 [J]. *Journal of Finance, 43(3)*, 541–566.

Rosenblatt, M. (1956). Remarks on some nonparametric estimates of a density function. *The Annals of Mathematical Statistics, 27(3)*, 832–837.

Shannon, C. E. (1948). A mathematical theory of communication. *Bell System Technical Journal, 27(3)*, 379–423.

Chapter 7
Genetic Algorithm-Based Combination of Predictions

Abstract We propose a function allowing us to combine predictions of the market, industry, and idiosyncratic components of the stock returns in a single portfolio. The combination is made through a double nested optimization, using numerical and genetic optimizations in order to take into account the transaction costs. The resulting aggregated trading strategy reaches a Sharpe ratio of 0.94, net of transaction costs, versus only 0.43 for the market portfolio. The genetic optimization proposed also outperforms a simple risk parity benchmark.

Keywords Genetic algorithm · Polymodel theory · Artificial intelligence · Machine learning · Alpha combination · Transaction costs · Cross-section of stock returns · Market timing · Trading strategy · Trading signal · Stock returns predictions · Risk parity

7.1 Introduction

The purpose of the present chapter is to propose a combination method for the predictions obtained in Chaps. 4, 5 and 6. These predictions are made by different means on the market, the industry, and the specific part of the stock returns. As presented in the introduction, the returns of a portfolio which aggregates such predictions can be expressed as a function of the returns of the sub-portfolios made with these different predictions:

$$r_P = \Omega_p(r_M, \mathcal{F}_I, \mathcal{F}_S). \tag{7.1}$$

Here Ω_p is the aggregation function of the sub-portfolios returns, r_M is the returns of the market factor portfolio, $\mathcal{F}_I$ is the returns of a portfolio composed with the set of the industry factors, and $\mathcal{F}_S$ is the returns of a portfolio composed with the set of the specific factors.

In our case, since we only have one industry and one specific factor in their respective sets, Eq. (7.1) can be simplified by simply setting:

© The Author(s), under exclusive license to Springer Nature Switzerland AG 2022
T. Barrau, R. Douady, *Artificial Intelligence for Financial Markets*, Financial Mathematics and Fintech, https://doi.org/10.1007/978-3-030-97319-3_7

$$r_P = \Omega_p(r_M, r_{AFI,I}, r_{PP,S}).\tag{7.2}$$

Here $r_{AFI,I}$ is the returns of the AntiFragility Indicator Industry portfolio, and $r_{PP,S}$is the returns of the Polymodel Predictions Specific portfolio.

Thus, the aim of this chapter is to propose a first, scalable definition of omega.

First, we note that omega is already partly defined by the signal elaborated in Chap. 4, dedicated to market timing. Indeed, since it is a function of the raw factor returns, omega includes the market and factor timing elements of the aggregated portfolio performance. But this point is of less importance, being just a matter of definition. More interestingly, we should observe that mixing cross-sectional predictions of the specific and industry returns with time-series predictions of the market returns does not result in an intuitive portfolio construction. Hence, since they are finally defined by positions taken on the same investment universe, we decide to directly aggregate portfolio positions (and not signals) of the different predictions, for the sake of simplicity.

Machine learning methods, and especially neural networks, have recently shown great improvements for the combination of predictors, compared to classical alternatives such as linear regression (as shown by Gu et al., 2018). However, since we only have three predictions to combine, defining omega in a non-linear manner seems to be an overkill. Hence, we choose to define omega by a linear weighting of the sub-portfolios.

Still, we shouldn't be over simplistic. Indeed, as is highlighted later in the present chapter, even in our simple case, the transaction costs matter a lot for the performance of the final implementation of the strategy. We thus incorporate into omega the management of the transaction costs in order to produce superior *net* returns.

The management of the transaction costs is made through partial execution of the trades, and appropriated weighting of the sub-portfolios. The weights associated to each of the sub-portfolios are determined by a genetic algorithm designed for this purpose. The main proposal of the current chapter is thus to use a genetic algorithm to maximize the net Sharpe ratio of a portfolio through an optimal combination of predictions of different components of the stock returns.

Genetic algorithms have already been used to combine different predictors. Zhang and Maringer (2016) proposed a genetic algorithm to combine cross-sectional predictors, that can be understood as factors. Although they did not take into account transaction costs in their algorithm design, the application of their genetic algorithm is still similar to ours. Allen and Karjalainen (1999) used a genetic algorithm mainly to discover technical trading rules (in/out the market signals based on predictors), but also to combine them. Again, the transaction costs are not introduced into their algorithm. In their 2012 paper, Sefiane & Benbouziane showed the efficiency of a genetic algorithm to construct a 5-asset portfolio, optimal in the mean-variance sense. Such a problem is also close to our current one, since we are looking to optimally combine a portfolio of three sub-portfolios, which may be considered as three assets. Sinha et al. (2015) also proposed a genetic algorithm for portfolio optimization, without transaction costs.

Genetic algorithms have also been used to limit transaction costs. Indeed, Lin and Wang (2002) proposed a genetic algorithm for mean-variance optimization with transaction costs, however their setting is slightly different (they use fixed transaction costs and round lots, while we use proportional transaction costs, which we consider to be more realistic).

Of course, the question of combining different predictors (or alphas) while reducing transaction costs has already been discussed in the seminal paper of Gârleanu and Pedersen (2013), who proposed to use dynamic programming to reduce trading fees. We re-use a part of their framework in the current chapter, which may be understood as an empirical implementation of the theoretical case mentioned in their part V ("Theoretical Applications"), Example 2 ("Relative-value trades based on security characteristics"). Apart from their study, a few other approaches have been developed to reduce transaction costs. For example, Ruf and Xie (2019) proposed a method to limit the impact of transaction costs on systematically generated portfolios, though they did not use a genetic algorithm for this purpose.

Thus, to the best of our knowledge, a genetic algorithm-based method which allows us to combine alphas regarding their performance net of transaction costs does not seem to exist in the literature, hence our proposal.

The rest of the chapter is organized as follows:

- In the first section we present the different predictions we are about to aggregate, with their definition, portfolio construction method and economic rationale.
- In the second section we propose a very simple combination method, standard in the literature, that is the benchmark for the genetic algorithm to outperform. We introduce here the question of the transaction costs along with a method "a la Gârleanu–Pedersen" to reduce them.
- In the third section we present the genetic algorithm used to weight the sub-portfolios corresponding to the different predictions.
- We finally perform a series of robustness tests and conclude the chapter.

7.2 Analysis of Strategies

7.2.1 Predictions of Market Returns

The predictions of market returns are those developed in Chap. 4. The signal used for the predictions is built by estimating a polymodel of the S&P 500. The RMSE of each of the elementary models is measured in order to form a distribution of RMSEs. This distribution captures the strength of the links the market maintains with its whole economic environment. Following a rolling method, the signal then learns the distribution of the RMSE that occurred before a bear phase. This distribution is compared to the current distribution to form a prediction of the direction of the market returns in the next month.

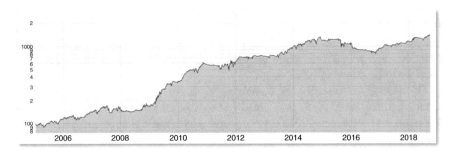

Fig. 7.1 Performance of the portfolio based on the predictions of market returns

Hence, the signal is an on/off investment signal which is implemented by investing in the market portfolio with a dynamic leverage, equal to 0.5 (short position) when the risk is off and +2 (long position) when the risk is on. The resulting portfolio is relatively slow, with a half-life[1] of 47 days. On average, the portfolio has a leverage of 1.37. Below is the cumulative performance[2] of this market timed portfolio, from 2005-01-01 to 2018-28-09 (Fig. 7.1). The Sharpe ratio over the period is 0.92.

7.2.2 Predictions of Industry Returns

The predictions of industry returns are those developed in Chap. 5. The signal used for the predictions is built using LNLM models with each industry return as the target variable, and the market returns as the explanatory variable. A measure of the fragility (in the sense of Taleb's antifragility concept (see Taleb, 2012)) is then computed, based on the concavity/convexity to the market of each industry. The signal then indicates the fragility of the industry, a large positive signal corresponding to a highly fragile industry. Indeed, fragile industries seem to benefit from a risk premium, since their exposure to extreme market deviations is compensated by superior returns.

The signal is then implemented in a long-short manner, expected to have a small market exposure on average. The positions of the portfolio are obtained using a characteristic portfolio (Grinold & Kahn, 2000), which is essentially a mean-variance optimization of the signal, implemented with a target volatility of 10%. The resulting portfolio is also a (very) slow portfolio, with a half-life of around 400 days, but, with an average leverage reaching 8.31, is more leveraged than the

---

[1]The half-life is the number of days needed to reach an autocorrelation of 50%. A half-life of 20 days means that the current portfolio has a 50% correlation with what were its positions 20 days before.

[2]Cumulative performance is computed with compounding, with a base 100 at the beginning of the period, and presented in log-scale.

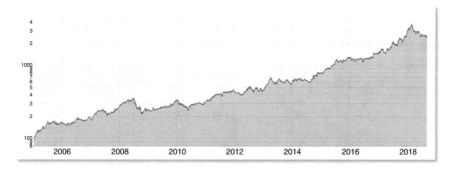

Fig. 7.2 Performance of the portfolio based on the predictions of industry returns

market portfolio. Note that such a level of leverage is less problematic for a long-short portfolio than for a long only portfolio, since the cost of the long position can be partly financed by the short position. Below is the cumulative performance of the industry portfolio, from 2005-01-01 to 2018-28-09 (Fig. 7.2). The Sharpe ratio achieved is 0.96.

7.2.3   Predictions of Specific Returns

The predictions of specific returns are those developed in Chap. 6. The signal is obtained from a straightforward application of polymodels: making predictions. For each company of the cross-section of stock returns, we estimate a polymodel using a large number of market variables (stock indices, currencies, commodities, etc...), the target variable being the specific returns of the company. This polymodel is then used to make predictions. After applying a selection and an aggregation method on the predictions, a combined prediction is obtained for each firm, thus forming a prediction for the entire cross-section of stock returns.

Similarly to the industry predictions, the signal is then implemented through a long-short characteristic portfolio with a target volatility of 10%. The resulting portfolio changes quickly, with a half-life of 9 days, and has a similar leverage to the industry portfolio (8.27 on average), due to the portfolio construction we choose. Below is the cumulative performance of the specific portfolio, from 2005-01-01 to 2018-28-09 (Fig. 7.3). The Sharpe ratio achieved is 1.07.

7.2.4   Correlation Analysis

An analysis of the correlation of the returns of the different portfolios among themselves, and with the market portfolio, allows us to understand the potential

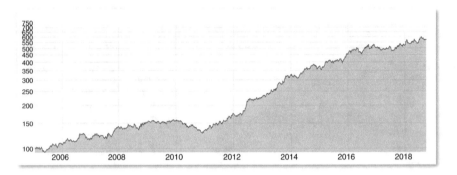

Fig. 7.3 Performance of the portfolio based on the predictions of specific returns

diversification benefit of combining them. Below is the correlation matrix of the daily returns (Fig. 7.4):

Unsurprisingly, the market portfolio (the benchmark) and the market timed portfolio are correlated, but the correlation is only 30%, which appears to be relatively low considering the method of implementation of the signal. The specific portfolio appears to be a great diversifier, while the industry predictions show some correlations to the market portfolio. These findings hold if we consider correlations using weekly returns, presented below (Fig. 7.5):

Using monthly returns, the market timed portfolio appears to be a bit more correlated to the benchmark, while the industry predictions are clearly less correlated (Fig. 7.6):

The relatively high (23%) correlation of the industry predictions we observed in the daily table (Fig. 7.4) is mainly explained by its tail correlation to the market portfolio. Indeed, keeping only the 1% most extreme market returns on both tails to compute the correlation leads to a tail correlation of 41% between the market and the industry portfolio. The industry portfolio is thus more correlated to the market during extreme moves. Since the returns of the industry portfolio come from an exposure to extreme market risk, it is expected that even a long/short portfolio construction is unable to entirely remove the market exposure.

Fig. 7.4 Correlation matrix of the daily returns of the trading strategies

	Benchmark	Industry Predictions	Market Predictions	Specific Predictions
Benchmark		23%	30%	7%
Industry Predictions	23%		9%	7%
Market Predictions	30%	9%		5%
Specific Predictions	7%	7%	5%	

Fig. 7.5 Correlation matrix of the weekly returns of the trading strategies

	Benchmark	Industry Predictions	Market Predictions	Specific Predictions
Benchmark		17%	35%	7%
Industry Predictions	17%		10%	7%
Market Predictions	35%	10%		6%
Specific Predictions	7%	7%	6%	

Fig. 7.6 Correlation matrix of the monthly returns of the trading strategies

	Benchmark	Industry Predictions	Market Predictions	Specific Predictions
Benchmark		7%	40%	8%
Industry Predictions	7%		-4%	12%
Market Predictions	40%	-4%		-11%
Specific Predictions	8%	12%	-11%	

In the aggregate portfolio we are about to construct, this negative feature is compensated by the negative tail correlations that the specific portfolio (-12%) and the market timed portfolio (-31%) exhibit with the benchmark.

7.3   Risk Parity Combination

7.3.1   Introducing Risk Parity

The combination of the three different portfolios may be done using 1/3 weights. Following this 1/N approach has been shown to be a robust method of aggregating predictions, difficult to beat in many cases (see Rapach et al., 2010; Timmermann, 2006). However, the predictions we used, even if considering the same assets, are distorted by different portfolio construction methods, making it more appropriate to aggregate sub-portfolios than predictions. Also, sub-portfolios are sufficiently uncorrelated to be considered as different assets. The 1/N method has also been

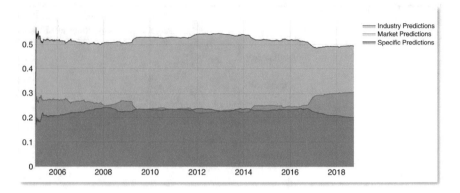

Fig. 7.7 Risk Parity combination: Strategies' weights over time

found to be robust when composing a portfolio of several assets (DeMiguel et al., 2009).

However, in our setting it is not clear what 1/N should be. Giving the same capital to portfolios with very different leverages, ranging from 1 to 8, seems unfair. A solution may be to scale the capital invested in the different portfolios, so that all the portfolios have the same amount of money effectively invested (which equals capital × leverage). But we directly encounter the problem that the market timing strategy is performing thanks to a dynamic leverage. A scaling of the capital thus does not seem appropriate to create a 1/N strategy.

An alternative is to give to each strategy the same volatility budget, leading to 1/volatility weights. This approach, called 'Risk Parity' has been implemented in various formats, both in academic works and empirical portfolios, and has shown good results (Clarke et al., 2013). We use it as a starting point to define an aggregation method benchmark. The weights obtained using a rolling 5-year window[3] to compute the volatility are as follows (Fig. 7.7):

We observe that the weights are very stable, which is an interesting property, since unstable weights would result in more turnover of the portfolio, and thus, more transaction costs. We also notice the relatively small volatility of the specific predictions portfolio, which results in a higher weight for the Risk Parity weighted portfolio.

Below is the performance over time of the aggregated portfolio obtained with this method (Fig. 7.8):

In addition to being very stable over time, the performance obtained showed very enviable metrics: the Sharpe ratio achieved is 1.58, an average annual return of 18%, 10% of realized volatility and a worst drawdown of 9%.

---

[3]The window is expanding at the beginning of the 2005–2018 period in order to avoid a rough truncation of the simulation.

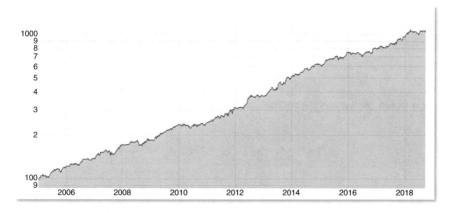

Fig. 7.8 Risk Parity portfolio gross performance

7.3.2   *Transaction Costs Matter*

We have presented a first, straightforward implementation of an aggregate portfolio, for which we saw a high *gross* performance. However, more than being a useful indicator of the predictive power of trading strategies, a simulated portfolio performance indirectly implies that the strategy may be profitable. Such a profitability, to be effective, must pass the final test of being implemented in the real world. Although this is out of the scope of the academic work we are conducting here, we can still reduce the distance between a real-world P&L and our simulation by introducing the transaction costs in our modeling.

The transaction costs of the equity market are composed of several elements: broker fees, different taxes, and of course the market impact of the trades. Proportional aggregated transaction costs, which encompass these different kinds of costs, have historically been found to be around 10 bips to 25 bips per unit of turnover[4] (Sweeney, 1988; Allen & Karjalainen, 1999). More recently Guo (2006) found a value of 25 bips, which seems to be in line with the latest studies (see for example Brière et al. (2019), who conducted a study using data from a large asset manager, making their results particularly relevant in our context of trading strategy implementation).

Being conservative, we choose to consider an amount of 25 bips of proportional transaction costs, which we added to the gross returns of the risk parity portfolio to form the net returns of the portfolio. Let us call "w" the vector of positions taken by the portfolio inside our investment universe of 500 stocks. At date t:

---

[4] x bips per unit of turnover should consider the leverage: if we invest $1 with a leverage of 8, the transaction costs for a round trip would be 8 × x, and not just x. This detail of implementation is important, considering the relatively high leverage that our portfolio may take.

$$w_t = (w_{1,t}, w_{2,t}, \ldots, w_{500,t}). \tag{7.3}$$

The trades "Δ" are given by the difference of the positions between two dates:

$$\Delta_t = w_t - w_{t-1} \tag{7.4}$$

and thus the portfolio performance net of transaction costs is:

$$r_{P,t} = \sum_{a=1}^{500} w_{a,t} \times r_{a,t} - |\Delta_{a,t}| \times 0.0025. \tag{7.5}$$

The performance is dramatically reduced by the addition of transaction costs, since the Sharpe ratio becomes equal to -4.88, which emphasizes the importance of the reduction of the gap between a theoretical and a real implementation of the strategy.

7.3.3 Ex-ante Optimal Reduction of the Transaction Costs

The impressive decrease of the performance observed while integrating the transaction costs can be greatly reduced by a less naive portfolio construction. Indeed, the trades are clearly inducing excessive costs.

Gârleanu and Pedersen (2013) popularized the usage of dynamic programming in finance. Their proposal to reduce the transaction costs of a portfolio is based on two core principles: aim in front of the target, and trade partially toward the aim. We focus here on the second principle, which is reflected in their paper by Eq. (7.6):

$$w_t = (1 - \theta)w_{t-1} + \theta \, aim_t. \tag{7.6}$$

Here w is the cross-section of the positions taken by the portfolio, aim is the portfolio we would like to achieve in a world without transaction costs, and θ is the optimal trading rate, i.e. the percentage of the total trades to the aim portfolio which maximizes the net returns.

We implement this principle by numerically computing the ex-ante optimal trading rate, using a 5-year rolling window (the window is again expanding at the beginning of the simulation). θ thus becomes a time-dependent parameter:

$$w_t = (1 - \theta_t)w_{t-1} + \theta_t \, aim_t. \tag{7.7}$$

Below is the value taken by θ over time (Fig. 7.9):

Although these values may seem small, considering the relatively high level of transaction costs (recall that the net Sharpe of the basic portfolio was -4.88), it is not

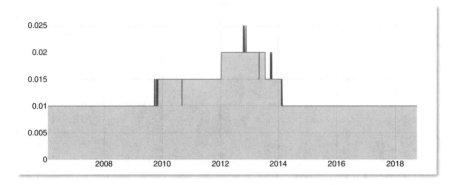

Fig. 7.9 Ex-ante optimal trading rate over time

Table 7.1 Annual net returns for different trading rates

Trading rate (%)	Average annual net return (%)
1.0	9.42
1.5	9.07
2.0	8.37
2.5	7.54
3.0	6.67
4.0	4.98
5.0	3.42

totally surprising. Moreover, higher values quickly lead to lower returns, as Table 7.1 shows:[5]

From this table, it is clear that exploring higher values of the trading rate is not worth it.

Using our ex-ante rolling optimal strategy the net returns reach 9.10%, which makes the results of the rolling method comparable with the best ex-post trading rate performances. These honest results should be considered together with the Sharpe ratio that we now achieve considering the net returns: 0.81. Below is the net cumulative performance of the signal (Fig. 7.10):

The performance achieved undoubtedly shows that an appropriate portfolio construction is essential to declare that the strategy may be profitable if implemented. As it has been previously shown in the literature (Ciliberti & Gualdi, 2020), the question of portfolio construction is thus an important determinant of the economic performance of a trading strategy.

---

[5] θs lower than 1% are not explored since lower values lead to unreliable misrepresentations of the original portfolio.

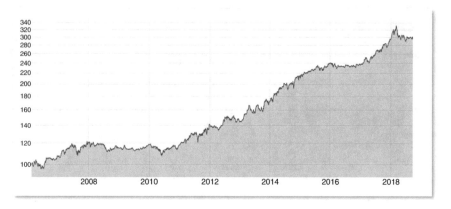

Fig. 7.10 Risk Parity portfolio net performance, with dynamic programming smoothing

7.4   Genetic Algorithm-Based Combinations

7.4.1   Methodology

We saw that Risk Parity allows us to reach a significant gross performance while combining portfolios, which can be partly preserved when transaction costs are introduced, by an important reduction of the amount traded. However, the presence of transaction costs raises uncertainty about the appropriateness of the choice of Risk Parity as a combination method. In particular, the analysis of the strategies showed that the predictions of specific returns correspond to the sub-portfolio that changes the most (by far), making it the potential principal contributor to the transaction costs of the aggregated portfolio. Parallelly, the specific portfolio is the one with the lower volatility, making a Risk Parity allocation to this portfolio of 50%. Thus, net of transaction costs, using Risk Parity may be a sub-optimal weighting choice.

A proper weighting algorithm should balance gross performance, transaction costs and risk, measured as portfolio's volatility. It would hence result in a superior net Sharpe ratio, which can be taken as a metric to optimize. Still, if we keep the rolling optimal trading rate mechanism presented before, the sub-portfolios weighting optimization problem would be nested in the trading rate optimization problem. The use of a numerical optimization algorithm may thus be appropriate to keep our final setting relatively simple.

Genetic algorithms would be a suitable candidate for this task. Introduced by Holland in 1975 (see Holland (1992) for a more recent version of the original paper), genetic algorithms have been widely applied in finance to various optimization problems (see Pereira (2000) for an introduction and a review of applications of genetic algorithms in finance). Being an optimization algorithm, it is indeed suitable for the sub-portfolios weighting problem.

Genetic algorithms represent the optimization problem in the form of a population of individuals (or chromosomes) which evolves while following mechanisms mimicking natural selection. We describe below how the population of chromosomes evolves until it reaches a final state, solving the optimization problem for date t, using a classical algorithm adapted to our current optimization problem.

The use of a genetic algorithm implies that the optimization is represented in the form of a set of competitive individuals. Each of these individuals is the representation of a solution of the optimization problem. An individual is characterized by its genotype, usually represented by a vector of 0s and 1s. In our case, the genotype of the vth individual is a vector which contains the weights of each sub-portfolio (positive and floored at 15% to keep a reasonable level of diversification):

$$g_v = \left(\beta_{v,M}, \beta_{v,I}, \beta_{v,S}\right) \quad s.t. \quad \beta_{v,M} + \beta_{v,I} + \beta_{v,S} = 1 \quad s.t. \quad \beta_{v,sub} \geq 0.15 \text{ for sub in } \{M, I, S\}. \tag{7.8}$$

Here $\beta_{v, M}$ is the weight of the Market sub-portfolio, $\beta_{v, I}$ is the weight of the (AntiFragility Indicator) Industry sub-portfolio, and $\beta_{v, S}$ is the weight of the (Polymodels Predictions) Specific sub-portfolio.

This genotype evolves over time, a dynamic simply represented as:

$$g_{v,t} = \left(\beta_{v,t,M}, \beta_{v,t,I}, \beta_{v,t,S}\right). \tag{7.9}$$

The algorithm starts by generating a random initial population "IP", i.e. a set of 50 individuals with genotypes randomly chosen in a uniform distribution:

$$IP_t = \left\{g_{1,t}, g_{2,t}, \ldots, g_{50,t}\right\}. \tag{7.10}$$

These individuals are then evaluated based on a fitness score. In our case, the fitness score of an individual is the net Sharpe ratio achieved by the portfolio weighted by its genotype, after trading rate optimization. At time t, the aggregated portfolio w, defined as the vector of the positions corresponding to the individual v is thus:

$$w_{v,t} = (1 - \theta_{v,t})w_{v,t-1} + \theta_{v,t}\left(\beta_{v,t,M} \times w_{t,M} + \beta_{v,t,I} \times w_{t,I} + \beta_{v,t,S} \times w_{t,S}\right). \tag{7.11}$$

Here, the optimal trading rate is computed every day, using the procedure previously described. Note that such a construction of the aggregated portfolio implies that the aggregation function omega is linear, depending on the trading rate:

$$\Omega_{v,t}(r_M, r_{AFI,I}, r_{PP,S}) \Rightarrow \Omega_{v,t}\left(\theta_{v,t}, \beta_{v,t,M}, \beta_{v,t,I}, \beta_{v,t,S}\right). \tag{7.12}$$

Hence, the realized net returns of the portfolio associated with individual v is:

$$r_{v,t} = \sum_{a=1}^{500} r_{a,t} \times w_{a,v,t} - |\Delta_{a,v,t}| \times 0.0025 \tag{7.13}$$

and so the fitness score of the individual is simply its Sharpe ratio, which we compute using the net portfolio returns over the 10 years that precede the current date:

$$S_{v,t} = \left(\sum_{s=t-252*10}^{t} r_{v,s} - risk\_free\_rate_s \right) / \sigma_{v,s \to t} \times \sqrt{252}. \tag{7.14}$$

Here $\sigma_{v, s \to t}$ is the standard deviation of the excess daily returns of the portfolio in the 10 years preceding date t.

Fitness scores are normalized so that they are all positive, by subtracting from all fitness scores the smallest of them:

$$S'_{v,t} = S_{v,t} - min(S_t). \tag{7.15}$$

The fitness score of an individual determines its reproduction probability for the next period, defined as:

$$rp_{v,t} = S'_{v,t} / \sum_{v=1}^{50} S'_{v,t}. \tag{7.16}$$

Then the reproduction phase starts, structured as follows:

- Two individuals, called "parents" are drawn randomly from the initial population, with a probability of being selected coming from Eq. (7.16) above.
- These two parents produce two children, which are obtained by an operation called "cross-over": the parents exchange a part of their genotype, which is 1 element (1/3 of the genotype) in our case, i.e. 1 sub-portfolio weight.
- Each of the children have a 30% probability of mutation. If one of the children mutates, a random genotype is given to it. This feature allows the population to maintain a genetic diversity, i.e. it prevents the optimization algorithm from converging too quickly.

The reproduction phase is repeated 25 times so that a new population "NP" composed of 50 individuals is generated. This new population will be the initial population of the next "epoch", which represents a new period of evolution of the population:

$$NP_{t,epoch=1} = IP_{t,epoch=2}. \tag{7.17}$$

This full process, which governs how an initial population evolves to become a new population, is then repeated 20 consecutive times, using the same data of the last 10 years available at date t:

$$IP_{t,epoch=1} \rightarrow NP_{t,epoch=1} = IP_{t,epoch=2} \rightarrow NP_{t,epoch=2} = \ldots = IP_{t,epoch=20}$$
$$\rightarrow NP_{t,epoch=20}. \tag{7.18}$$

For example, starting on December 31st, 2015, we use the data from January 1st, 2006 to December 31th, 2015. We create an initial random genotype and apply the reproduction/selection procedures as described above, updating the initial population at each epoch by making it the new population of the previous epoch. We finally average the genotypes of the last generation of individuals. At this stage, the diversity of the population has decreased because the selection process tends to eliminate non-fit individuals over time.

The average genotype is thus given by the final sub-portfolio weights we obtained at date t. These weights are then used to produce the returns of the final aggregated portfolio we consider over the next year. Returning to the previous example, we used the data available until 31st December 2015 to produce sub-portfolio weights. The aggregated portfolio performance obtained in this way is thus only of interest in 2016 and later (otherwise the performance presented suffers from in-sample optimization). We consider this performance only for the year 2016, after which the full process is repeated to generate the aggregated portfolio's out-of-sample performance of 2017, and so on, following this rolling methodology.[6]

7.4.2   Results

Below are the sub-portfolio weights produced by the genetic algorithm (Fig. 7.11):

As expected, since the fitness score of the individuals is based on their *net* returns, the algorithm tends to under-weight the specific portfolio compared to risk parity, since this portfolio may be very detrimental in terms of transaction costs. Apart from this point, we can see that even if the weights are changing over-time, there is no extreme recombination of the sub-portfolios weighting scheme, even if it seems that the market portfolio tends to be more invested as time passes.

The net Sharpe ratio of the strategy is 0.94 versus 0.81 for the Risk Parity portfolio, while the annual net return is 10.40%, versus 9.10% in the Risk Parity case. Below are the two cumulative P&L curves (Fig. 7.12):

Starting from a value of 100 in 2006-01, the final value reached by both portfolios on 2018-10 is 353 for the genetic algorithm combined portfolio and 295 for the Risk Parity combined portfolio. The genetic algorithm portfolio seems to deliver higher

---

[6]Hence the optimal θ is recomputed every day, while the sub-portfolio weights are only computed once a year, a choice mainly motivated by computational limitations.

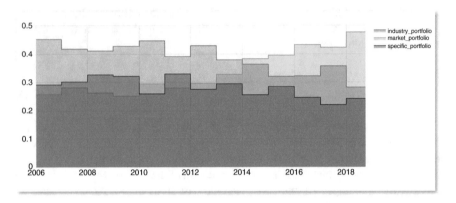

Fig. 7.11 Genetic algorithm combination: Strategies' weights over time

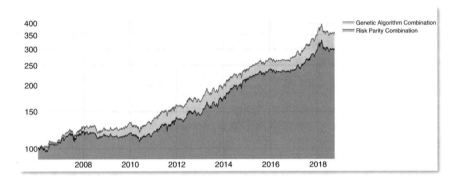

Fig. 7.12 Comparison of performances of Risk Parity and genetic algorithm combinations

returns most of the time, as one can assess from the cumulative P&L difference of both strategies (Fig. 7.13):

On top of this economically significant increase of the performance, the improvement is also statistically significant. This can be evaluated by modeling the genetic portfolio returns using a linear model, including a constant and the Risk Parity returns. Estimated with OLS, the constant parameter of such a model shows a t-stat of 2.30, corresponding to a 2.2% p-value.

The portfolio aggregated with the genetic algorithm is also over-performing the market portfolio. Below are the P&L curves of both portfolios (standardized at 10% annual volatility to be comparable) (Fig. 7.14):

The final aggregated portfolio shows a correlation of 40% with the market portfolio. However, the Sharpe ratio[7] of the genetic portfolio reaches 0.94 over the period, versus only 0.43 for the market portfolio. If we consider the worst

---

[7]Computed on net returns, assuming that the market portfolio can be invested at no costs.

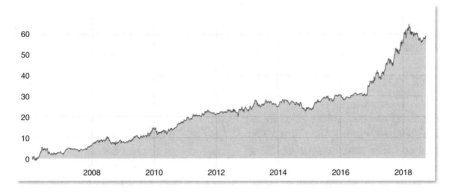

Fig. 7.13 Cumulative P&L difference over time between combination methods

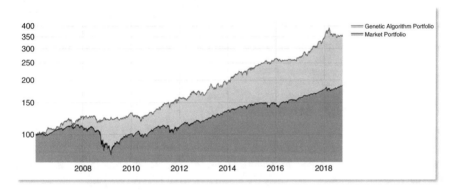

Fig. 7.14 Comparison of performances of the genetic algorithm portfolio and the market portfolio

drawdown[8] reached by the market portfolio, it is -55% for the market portfolio, while being only -22% for the genetic portfolio.

7.5   Robustness Tests

We provide a sensitivity analysis of the strategy's performance with the different parameters used by the genetic algorithm.

---

[8]Computed with returns standardized to have the same full sample volatility as the market portfolio. For a 10% full sample volatility, the worst drawdown of the genetic portfolio would be -12%.

Table 7.2 Sensitivity of the Sharpe ratio to the mutation probability

Mutation probability	10%	20%	30%	40%	50%
Net sharpe ratio	0.91	0.92	0.94	0.91	0.92

7.5.1 Mutation Probability

The mutation probability used in the version of the algorithm presented above is 30%. Below, we display the net Sharpe ratio of the strategy for 10%, 20%, 30%, 40% and 50% mutation probabilities:

The performance seems to be relatively robust to the change in this parameter. It seems to reach an optimum at 30%, a high level that make sense as the mutation probability is the main defense against overfitting of the algorithm. For our given number of epochs, a lower probability results in a potential early convergence to "extreme" solutions, while increasing the probability progressively prevents the convergence of the algorithm. Note that support for this reasoning comes more from its internal validity than by inspecting the figures of Table 7.2, as we can cast some doubts about the statistical significance of Sharpe ratio differences of 0.02.

7.5.2 Number of Chromosomes

In the version of the algorithm presented above we use 50 chromosomes. Below, we display the net Sharpe ratio of the strategy for 20, 35, 50, 75 and 100 chromosomes (Table 7.3):

A higher number of chromosomes may provide more robust results, but it is also at the cost of a higher computation time, which does not seem justified by the performance. The performance achieved with 75 chromosomes seems a bit worrying. An empirical implementation of the strategy should take this point into account, for example by averaging the sub-portfolio weights obtained with different numbers of chromosomes.

7.5.3 Number of Epochs

In the version of the algorithm presented above we use 20 epochs. Below, we display the net Sharpe ratio of the strategy for 5, 10, 20, 30 and 50 epochs (Table 7.4):

Table 7.3 Sensitivity of the Sharpe ratio to the number of chromosomes

Number of chromosomes	20	35	50	75	100
Net sharpe ratio	0.92	0.92	0.94	0.89	0.92

Table 7.4 Sensitivity of the Sharpe ratio to the number of epochs

Number of epochs	5	10	20	30	50
Net sharpe ratio	0.88	0.92	0.94	0.93	0.91

Table 7.5 Sensitivity of the Sharpe ratio to the randomness seed

Seed	#1	#2	#3	#4	#5
Net Sharpe Ratio	0.94	0.92	0.93	0.94	0.89

Table 7.6 Sensitivity of the Sharpe ratio to the optimal trading rate window

Optimal trading rate window	3 years	4 years	5 years	6 years	7 years
Net sharpe ratio	0.92	0.92	0.94	0.92	0.93

Here, we see again that the parameter selected corresponds to an optimum, related to the questions of over and under-fitting. It is plausible that other configurations of the parameters, if adjusted jointly, may lead to a similar or superior performance (for example, with a larger mutation probability and more epochs).

7.5.4   Seed

The seed[9] used to get replicable randomness in the version of the algorithm presented above is labeled #1. Randomness is used for selecting mutating chromosomes, generating initial population weights and selecting the fittest individuals for reproduction. Below, we display the net Sharpe ratio of the strategy for the seeds #1, #2, #3, #4 and #5 (Table 7.5):

Seeds other than the one selected deliver comparable performances, except for seed #5. This justifies the need to average sub-portfolio weights obtained from different seeds to avoid seed sensitivity of the performance.

7.5.5   Optimal Trading Rate Window

The genetic algorithm portfolio takes into account the transaction costs thanks to the usage of an ex-ante optimal trading rate, which maximizes the net Sharpe ratio measured on a given rolling window. The length of the rolling window used in the version of the algorithm presented above is 5 years. Below, we display the net Sharpe ratio of the strategy for 3, 4, 5, 6 and 7 years (Table 7.6):

Again the performance seems to be quite insensitive to a change in this parameter.

---

[9]The seed is provided by the python library Numpy, with np.random.seed(1).

Overall, the performance of the aggregated strategy obtained using a genetic algorithm-based combination does not particularly suffer from adopting other configurations of the algorithm's parameters. Such a behavior indicates a high level of robustness of the results presented above.

7.6   Conclusions

The present chapter shows that the portfolio construction is of primary importance in the implementation of a trading strategy. It proposes a solution to the problem of weighting sub-portfolios reflecting different predictions of the stock returns while significantly reducing the transaction costs. Such a solution is non-trivial since it involves a double-nested optimization problem.

Other optimization methods may be explored, for example the (stochastic) gradient descent, given its success in other machine learning applications such as neural networks.

Still, the results presented show that genetic algorithms are suitable for the alpha combination problem, allowing us to perform alpha combinations while efficiently taking the transaction costs into account. Such a portfolio construction method seems to add value on top of a classical benchmark, and to the best of our knowledge it has not already been proposed in the literature. Moreover, the aggregated portfolio exhibits twice the Sharpe ratio of the market in the period considered, associated with a reduced correlation to the market and a significant reduction of the drawdowns.

From the perspective of the aggregate predictions, this shows that the resulting trading strategy may be profitable. Indeed, the current setting is fairly realistic, thanks to the inclusion of transaction costs. Nevertheless, the realism of the simulation presented is still imperfect, since the only way to assess the profitability of a trading strategy is to implement it (this is the only solution in which all the possible biases, such as look-forward bias, selection bias, technical biases (. . .) are removed). However, we expect that the gap to such a level of realism has been partly filled by the inclusion of transaction costs. Note that we indirectly take into account the financing of the positions since the risk-free rate is deducted from the returns in the computation of the Sharpe ratios presented. Leverage costs are nevertheless not included and may be considered in a further study.

References

Allen, F. & Karjalainen, R. (1999). Using genetic algorithms to find technical trading rules. *Journal of Financial Economics, 51(2)*, 245vf271.

Brière, M., Lehalle, C. A., Nefedova, T., & Raboun, A. (2019). Stock market liquidity and the trading costs of asset pricing anomalies. Available at SSRN 3380239.

Ciliberti, S., & Gualdi, S. (2020). Portfolio Construction Matters. *The Journal of Portfolio Management, 46*(7), *46–57.*

Clarke, R., De Silva, H., & Thorley, S. (2013). Risk parity, maximum diversification, and minimum variance: An analytic perspective. *The Journal of Portfolio Management, 39*(3), 39–53.

DeMiguel, V., Garlappi, L. & Uppal, R. (2009). Optimal versus naive diversification: How inefficient is the 1/N portfolio strategy? *The Review of Financial Studies, 22*(5), 1915vf1953.

Gârleanu, N., & Pedersen, L. H. (2013). Dynamic trading with predictable returns and transaction costs. *The Journal of Finance, 68*(6), 2309–2340.

Grinold, R. C., & Kahn, R. N. (2000). *Active portfolio management: A quantitative approach for producing superior returns and controlling risk.* McGraw-Hill.

Gu, S., Kelly, B., & Xiu, D. (2018). *Empirical asset pricing via machine learning* (No. w25398). National Bureau of Economic Research.

Guo, H. (2006). On the out-of-sample predictability of stock market returns. *The Journal of Business, 79*(2), 645–670.

Holland, J. H. (1992). *Adaptation in natural and artificial systems: an introductory analysis with applications to biology, control, and artificial intelligence.* MIT Press.

Lin, D., & Wang, S. (2002). A genetic algorithm for portfolio selection problems. *Advanced Modeling and Optimization, 4*(1), 13–27.

Pereira, R. (2000). *Genetic algorithm optimization for finance and investments.* MPRA Paper 8610, University Library of Munich.

Rapach, D. E., Strauss, J. K., & Zhou, G. (2010). Out-of-sample equity premium prediction: Combination forecasts and links to the real economy. *The Review of Financial Studies, 23*(2), 821–862.

Ruf, J., & Xie, K. (2019). The impact of proportional transaction costs on systematically generated portfolios. *arXiv preprint arXiv:1904.08925.*

Sefiane, S., & Benbouziane, M. (2012). Portfolio selection using genetic algorithm. *Journal of Applied Finance & Banking, 2,* 143–154.

Sinha, P., Chandwani, A., & Sinha, T. (2015). Algorithm of construction of optimum portfolio of stocks using genetic algorithm. *International Journal of System Assurance Engineering and Management, 6*(4), 447–465.

Sweeney, R. J. (1988). Some new filter rule tests: Methods and results. *Journal of Financial and Quantitative Analysis, 23*(3), 285–300.

Taleb, N. N. (2012). *Antifragile: Things that gain from disorde.* Random House Trade Paperbacks.

Timmermann, A. (2006). Forecast combinations. *Handbook of Economic Forecasting, 1,* 135–196.

Zhang, J., & Maringer, D. (2016). Using a genetic algorithm to improve recurrent reinforcement learning for equity trading. *Computational Economics, 47*(4), 551–567.

Chapter 8
Conclusions

Abstract We conclude the book by giving a summary of the answers it brings to the different questions raised at the beginning. We recap how the different concerns about the limits of Polymodel Theory have been addressed. Returning to the different applications discussed throughout the book, we outline how polymodels successfully contribute to the literature of financial market predictions, thus providing an effective artificial intelligence technique when applied to financial markets.

Keywords Polymodel theory · Artificial intelligence · Machine learning · Genetic optimization · Market timing · Cross-section of stock returns · Cross-section of industry returns · Trading strategy · Trading signal · Stock returns predictions

In the introduction of the book we presented a simple model of portfolio returns, which is represented as a function of the market, the industry and the specific components of the stock returns:

$$r_P = \Omega_p(r_M, \mathcal{F}_I, \mathcal{F}_S). \tag{8.1}$$

Still in the introduction, we emphasized the fact that superior portfolio performances, which add value on top of the simple portfolios such as the market portfolio, may come from three sources:

- Identifying unknown factors, known as "alphas", which explains industry or specific returns.
- Timing the performance of the sub-portfolios, through factor or market timing.
- Properly assembling the different sub-portfolios, i.e. defining the weighting in omega.

In Chap. 4, we proposed a market timing signal which seems to be an effective predictor of future market returns. This is a contribution to the second source of superior portfolio performances. The chapter contributes to the broad literature on the detection of Systematic Risk via the analysis of correlation structures of the stock markets. Instead of measuring correlations, we estimated a polymodel and measured the root-mean squared error of each of the elementary models. The distribution of the

---

RMSE was found to be predictive of future market drawdowns in a non-trivial manner, since we observed an extremization of the links, some economic variables being more correlated to the market before crisis, and some others (most of them) less. Such an application of Polymodel Theory shows its effectiveness as a tool to analyze a particular node of a complex system. We derived a market timing strategy from our Systemic Risk indicator, which had a gross Sharpe ratio of 0.68 for the S&P 500, versus 0.35 for being simply invested in the market (these numbers are computed with equal volatility of portfolio returns). This increase in the market portfolio performance is shown to be highly statistically significant, and subsists despite a large panel of robustness tests.

We contributed to the first source of over-performance in Chap. 5 by proposing a new industry risk premium, the antifragility factor, derived from the concept of antifragility developed by Taleb. It has been shown to be a significant predictor of industry returns, particularly profitable (the gross Sharpe ratio achieved is 1.10). The signal survives a large array of robustness tests, and regressions on well-known factors show that the alpha is different than the classical Fama & French factors. Notably, it subsumes other factors that try to capture the premia for extreme market risk (Downside Beta and Coskewness). Thus, this chapter is another contribution to the literature which shows the importance of non-linear modeling in finance.

Another contribution to this first source of superior portfolio performance is brought by Chap. 6, which uses polymodels as a machine learning method to predict the specific component of the cross-section of stock returns. For each of the 500 stocks of the cross-section we invest in, we define a polymodel of 1,134 predictors, which is re-estimated dynamically. Sophisticated selection and aggregation methods are developed to tackle this datamining exercise. The selection is performed using a rolling double filter and the aggregation takes into account both the goodness of fit and the correlation of the predictors thanks to the use of Information Theory. The usage of a prediction aggregation method which takes into account the correlation of the predictors is an important question for Polymodel Theory (as mentioned in Chap. 2), which finds a first answer in this chapter. The trading strategy implemented on the base of these specific predictions showed a gross Sharpe ratio of 0.91.

The third source of portfolio over-performance consists in defining an appropriate combination function "omega" for market, industry and specific predictions. This problem is tackled in Chap. 7, in which we combine the factor portfolios obtained from the three previous chapters. Using a Risk Parity combination, the aggregate trading strategy reached a gross Sharpe ratio of 1.58, which shows the high and consistent predictive power of the predictions. While introducing realistic (potentially conservative) values of transaction costs, the performance declines, justifying the development of a combination method based on a genetic algorithm, partly inspired by the dynamic programming model of Gârleanu and Pedersen. The genetic algorithm aims to solve a nested optimization problem which allows us to simultaneously obtain optimal weights for the three sub-portfolios as well as the optimal trading rate for the strategy. It is shown to outperform the Risk Parity combination from the statistical significance perspective. The final aggregated portfolio has a

Sharpe ratio net of transaction costs of 0.94. A natural benchmark for this portfolio would be the simple portfolio defined in the introduction:

$$r_{simple\_portfolio} = \beta_M r_M + \beta_{mom} r_{mom} + \beta_{size} r_{size} + \beta_{value} r_{value}. \tag{8.2}$$

Since our aim was to produce novel alpha signals and methods, we avoided any investment in the momentum, size and value factors. Hence, the portfolio against which the final strategy is benchmarked is the market portfolio, which reached a Sharpe ratio of 0.43 on the same period. Thus, net of transaction costs, the aggregated trading strategy more than doubles the performance of the market portfolio, which is the most convincing indicator of the predictive power of the predictions developed throughout the book.

The book thus contributes to the three sources of superior portfolio performance mentioned in the introduction. The development of these novel prediction methods and alpha signals would not have been possible without Polymodel Theory, which is extensively presented in Chap. 2. Complementing the empirical applications of polymodels made in the following chapters, this second chapter highlights the theoretical benefits of using polymodels for the general purpose of modeling. In particular, it shows that polymodels are especially suitable for solving artificial intelligence problems which need to manage large amounts of data in a data-driven, robust, accurate, and non-overfitted manner.

For the particular question of the estimation of the elementary models which compose the polymodels, a novel technique, the Linear-Non-Linear-Mixed model, is presented in Chap. 3. The LNLM model proposes to regularize the non-linear part of a polynomial model by mixing it with a linear model. The resulting shrinkage of non-linearity is obtained by a customized usage of Stratified Cross-Validation. Simulations showed both the effectiveness of the LNLM model to reduce over-fitting as well as good performances in terms of computation time. This last point is particularly important since some of the applications presented in the book required several billion model estimations. LNLM was thus revealed to be an appropriate solution to work in a big data framework, on top of being successfully used to produce the predictions of Chaps. 4, 5 and 6.

Polymodels estimated by the LNLM model are shown throughout the book to be useful for financial applications, however, these tools are quite general, and can be used in any field in which applied mathematics find interest. The prediction methods, as they ultimately capture behavioral biases which prevent market efficiency, also have the potential to be applied in any financial market and are not restricted to the equity market.

The simple model we presented for describing the returns of a portfolio invested in the stock market should be enriched. If fed properly by incorporating the vast number of predictors which exist in the literature, it can become an extremely powerful framework for constructing predictions. The aggregation function omega is scalable in the form presented in the book, but it should evolve in a more general direction, for example by being estimated through a deep neural network in order to benefit from the latter's property of being a universal function approximator.

The results of the book again confirm the absence of market efficiency. We have proved the possibility of consistently beating the market, but bearing in mind the perspectives that the framework we established offers, the book is clearly only a first draft of what is achievable regarding the prediction of financial markets.

These results strongly support the usefulness of Polymodel Theory among other artificial intelligence methods for the general sake of modeling.

Appendix

Representative RMSE Distributions per Stock Index

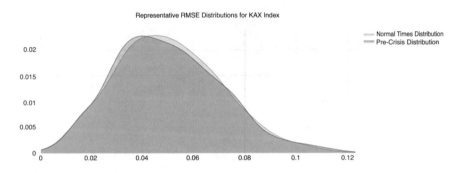

Representative RMSE Distributions for KAX Index

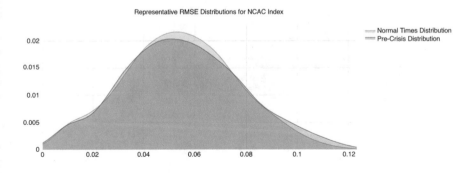

Representative RMSE Distributions for NCAC Index

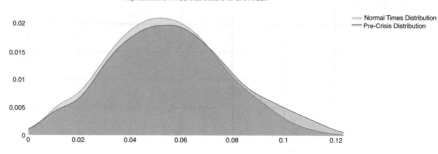

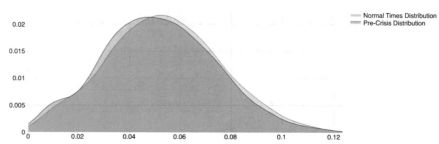

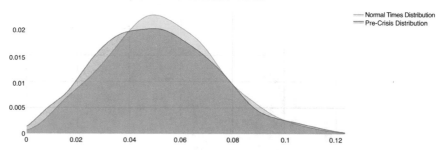

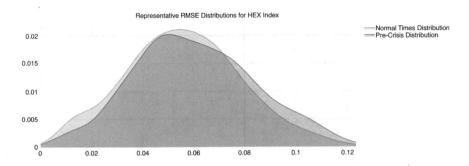

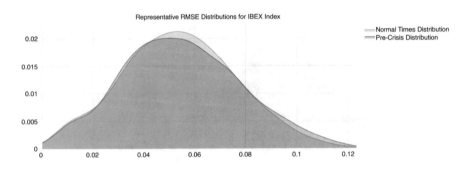

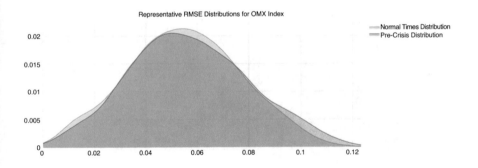

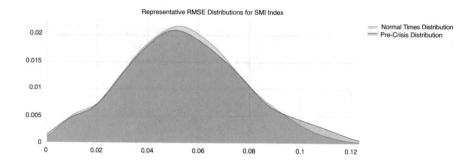

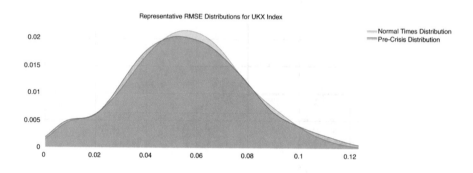

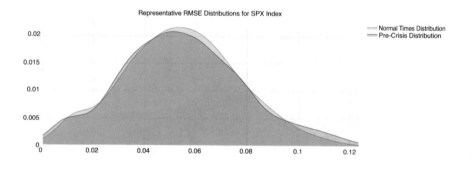

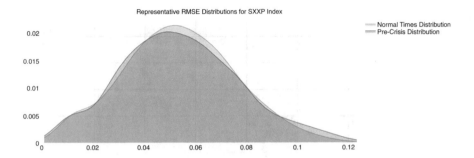

Market Timed Portfolio and Systemic Risk Indicator per Stock Index

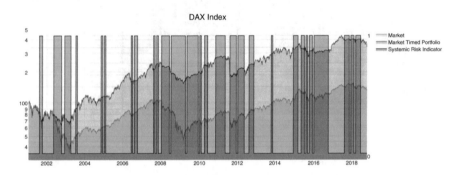

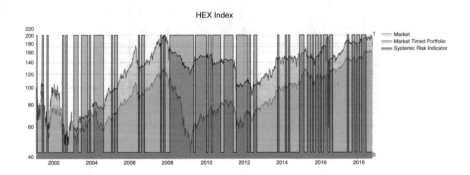

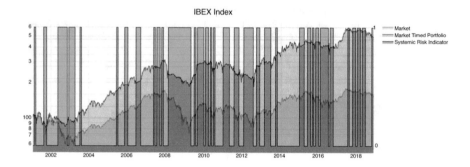

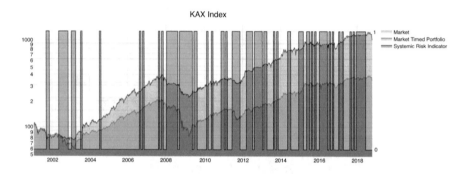

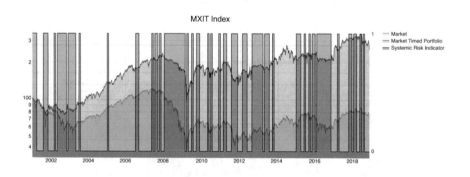

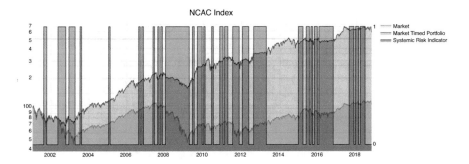

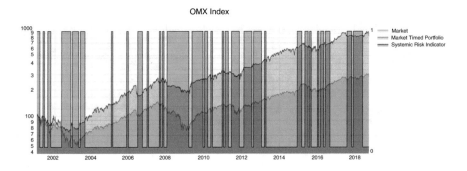

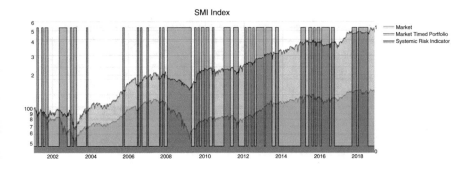

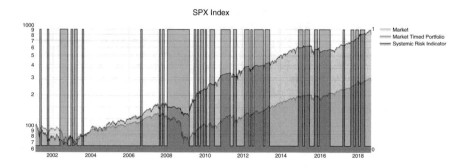

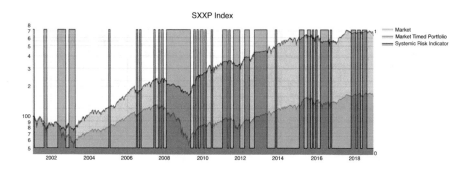

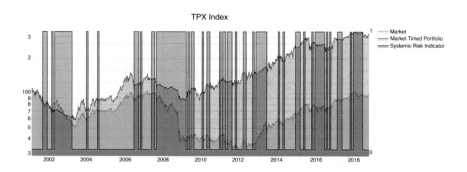

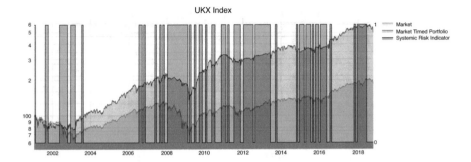

UKX Index

Industry Buckets Summary Statistics

Summary Statistics of Industry Buckets: Counts of sub-elements, per Industry

	Mean	Std	Min	Median	Max
101010	9.54	3.59	3	10	16
101020	32.28	10.92	15	34	52
151010	12.48	2.91	7	13	18
151020	1.37	0.52	1	1	4
151030	2.11	1.33	1	2	6
151040	5.55	1.87	1	5	11
151050	2.7	1.53	1	2	7
201010	10.73	1.63	6	11	13
201020	1.76	0.6	1	2	4
201030	2.06	1.3	1	2	6
201040	4.69	1.64	1	4	11
201050	3.44	0.67	2	3	5
201060	11.94	2.02	7	12	18
201070	2.03	0.66	1	2	4
202010	6.25	2.87	3	5	14
202020	2.42	1.41	1	2	6
203010	3.41	1	1	4	5
203020	2.61	1.28	1	3	6
203040	4.67	0.82	3	4	6
251010	3.23	1.66	1	3	8
251020	3.33	0.7	1	3	4
252010	6.43	3.36	1	6	15
252020	1.86	0.58	1	2	4
252030	4.66	1.76	1	4	9
253010	12.93	1.87	7	13	17
253020	2.18	0.97	1	2	5
254010	21.52	4.83	12	21	33
255010	1.27	0.44	1	1	2

(continued)

Summary Statistics of Industry Buckets: Counts of sub-elements, per Industry

	Mean	Std	Min	Median	Max
255020	4	1.29	1	4	6
255030	7.86	2.44	3	8	16
255040	14.92	1.79	11	15	20
301010	7.63	1.14	5	8	10
302010	6.59	1.42	4	6	9
302020	13.47	1.42	11	13	16
302030	3.23	0.79	1	3	5
303010	4.6	0.57	4	5	6
303020	2.36	0.69	1	2	4
351010	13.68	2.52	8	14	19
351020	15.14	3.22	5	15	22
351030	1.19	0.39	1	1	2
352010	8.48	2.09	2	8	14
352020	12.14	2.26	7	12	16
352030	4.07	1.63	1	5	6
401010	19.43	7.03	7	18	33
401020	6.45	3.32	1	6	12
402010	7.13	2.61	2	8	13
402020	4.28	0.74	3	4	6
402030	16.32	3.13	11	15	25
403010	24.33	4.3	16	24	33
404020	17.39	6.77	3	16	32
451010	6.02	3.49	1	5	17
451020	12.16	2.84	7	12	20
451030	15.19	3.92	7	16	28
452010	7.47	3.27	4	6	20
452020	8.84	2.08	4	9	14
452030	4.99	1.97	2	4	11
452040	1.05	0.21	1	1	2
453010	17.68	3.64	13	17	33
501010	5.62	2.42	3	5	13
501020	3.42	1.44	1	3	7
551010	14.93	5.48	10	12	34
551020	2.86	1.87	1	2	7
551030	8.28	3.83	1	10	17
551040	1	0	1	1	1
551050	3.25	1.3	1	4	6
601020	1.04	0.19	1	1	2

Summary Statistics of Industry Buckets: Market Capitalization

	Mean	Std	Min	Median	Max
101010	1.19E+11	5.81E+10	2.11E+10	1.00E+11	2.82E+11
101020	1.05E+12	4.32E+11	3.12E+11	1.08E+12	2.00E+12
151010	2.66E+11	1.30E+11	9.05E+10	2.18E+11	5.99E+11
151020	1.12E+10	8.95E+09	2.95E+09	7.01E+09	3.37E+10
151030	1.71E+10	1.89E+10	2.56E+09	8.95E+09	7.00E+10
151040	9.15E+10	4.37E+10	1.90E+10	8.90E+10	2.50E+11
151050	3.06E+10	1.35E+10	4.09E+09	3.10E+10	6.32E+10
201010	3.20E+11	1.47E+11	1.07E+11	3.00E+11	7.30E+11
201020	1.60E+10	7.74E+09	2.68E+09	1.60E+10	5.83E+10
201030	1.59E+10	1.02E+10	2.89E+09	1.30E+10	5.08E+10
201040	7.30E+10	2.87E+10	2.19E+10	7.09E+10	1.31E+11
201050	3.93E+11	9.38E+10	1.02E+11	4.15E+11	6.40E+11
201060	1.98E+11	8.50E+10	6.25E+10	2.05E+11	4.24E+11
201070	1.88E+10	1.31E+10	3.19E+09	1.35E+10	5.40E+10
202010	6.26E+10	1.97E+10	3.08E+10	6.01E+10	1.18E+11
202020	2.40E+10	2.35E+10	2.03E+09	1.20E+10	9.57E+10
203010	1.05E+11	5.15E+10	6.54E+09	1.18E+11	2.25E+11
203020	4.05E+10	4.06E+10	6.45E+09	2.14E+10	1.42E+11
203040	1.03E+11	5.78E+10	2.75E+10	9.57E+10	2.76E+11
251010	3.64E+10	2.46E+10	5.16E+09	2.77E+10	1.03E+11
251020	9.30E+10	4.60E+10	4.07E+09	8.51E+10	1.75E+11
252010	5.04E+10	2.84E+10	6.30E+09	4.40E+10	1.19E+11
252020	1.66E+10	5.97E+09	6.15E+09	1.57E+10	4.10E+10
252030	8.07E+10	6.11E+10	4.82E+09	5.88E+10	2.31E+11
253010	2.54E+11	1.28E+11	7.80E+10	2.35E+11	5.72E+11
253020	1.57E+10	7.51E+09	2.72E+09	1.58E+10	3.88E+10
254010	5.25E+11	1.42E+11	1.68E+11	5.28E+11	7.91E+11
255010	1.20E+10	6.95E+09	4.00E+09	8.27E+09	2.90E+10
255020	1.90E+11	2.40E+11	4.72E+09	8.58E+10	1.28E+12
255030	1.64E+11	1.03E+11	4.66E+10	1.19E+11	6.08E+11
255040	2.90E+11	1.01E+11	1.19E+11	2.57E+11	5.98E+11
301010	3.59E+11	1.48E+11	8.03E+10	3.65E+11	6.37E+11
302010	3.14E+11	8.34E+10	1.83E+11	2.78E+11	5.26E+11
302020	2.21E+11	8.59E+10	9.48E+10	2.08E+11	4.64E+11
302030	1.91E+11	8.84E+10	4.81E+10	1.77E+11	4.36E+11
303010	2.59E+11	6.93E+10	6.08E+10	2.60E+11	3.86E+11
303020	4.19E+10	1.68E+10	1.04E+10	4.06E+10	8.90E+10
351010	2.91E+11	1.46E+11	9.64E+10	2.63E+11	8.36E+11
351020	2.92E+11	1.59E+11	3.53E+10	2.56E+11	8.50E+11
351030	1.65E+10	9.04E+09	2.79E+09	1.64E+10	3.69E+10
352010	2.95E+11	1.93E+11	3.35E+10	2.23E+11	7.73E+11
352020	8.76E+11	1.62E+11	5.28E+11	8.66E+11	1.35E+12

(continued)

Summary Statistics of Industry Buckets: Market Capitalization

	Mean	Std	Min	Median	Max
352030	6.98E+10	5.42E+10	2.47E+09	5.08E+10	2.32E+11
401010	6.47E+11	3.61E+11	8.35E+10	5.40E+11	1.72E+12
401020	1.24E+11	8.12E+10	4.95E+09	1.59E+11	2.37E+11
402010	4.24E+11	1.64E+11	3.12E+10	4.26E+11	9.73E+11
402020	1.26E+11	3.83E+10	1.75E+10	1.23E+11	2.50E+11
402030	4.20E+11	1.59E+11	1.49E+11	3.82E+11	9.65E+11
403010	5.44E+11	1.12E+11	2.40E+11	5.30E+11	8.48E+11
404020	2.41E+11	1.95E+11	1.50E+10	1.65E+11	6.74E+11
451010	3.98E+11	3.97E+11	7.10E+09	2.48E+11	1.74E+12
451020	3.45E+11	2.80E+11	6.79E+10	1.65E+11	1.25E+12
451030	6.52E+11	3.00E+11	2.98E+11	5.51E+11	1.93E+12
452010	3.16E+11	1.87E+11	1.10E+11	2.69E+11	1.30E+12
452020	5.88E+11	1.97E+11	2.20E+11	5.51E+11	1.20E+12
452030	5.60E+10	2.58E+10	6.61E+09	5.39E+10	1.36E+11
452040	1.24E+10	6.33E+09	3.14E+09	1.20E+10	4.30E+10
453010	4.21E+11	1.90E+11	1.40E+11	3.63E+11	1.14E+12
501010	3.78E+11	1.11E+11	2.14E+11	3.55E+11	7.41E+11
501020	5.06E+10	2.03E+10	2.23E+09	5.17E+10	9.65E+10
551010	2.39E+11	7.70E+10	1.25E+11	2.07E+11	4.58E+11
551020	1.91E+10	1.39E+10	4.64E+09	1.34E+10	7.23E+10
551030	1.20E+11	6.30E+10	7.05E+09	1.27E+11	2.33E+11
551040	8.86E+09	4.12E+09	2.64E+09	8.58E+09	1.65E+10
551050	3.62E+10	2.31E+10	6.48E+09	2.73E+10	9.71E+10
601020	9.13E+09	3.34E+09	4.19E+09	8.67E+09	1.70E+10

AFI Scores per Industry over Time

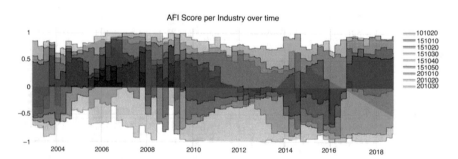

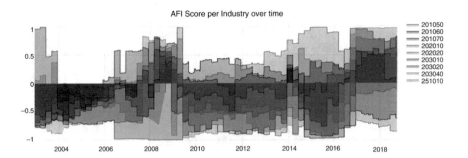

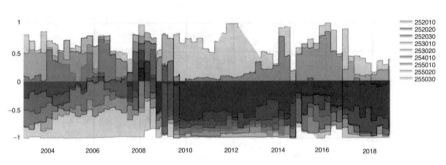

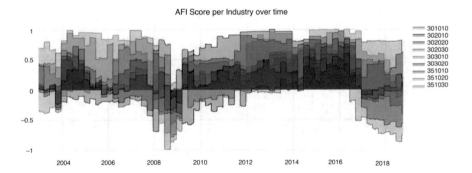

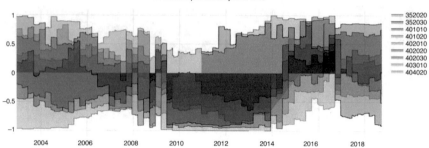

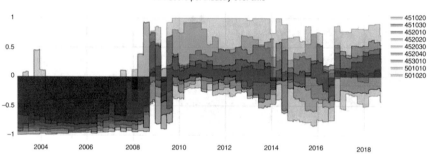

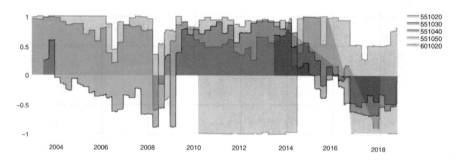

Printed in the United States
by Baker & Taylor Publisher Services